Rating Law and Valuation

RATING LAW AND VALUATION

Frances Plimmer

 LONGMAN

Pearson Education Limited
Edinburgh Gate
Harlow
Essex CM20 2JE
England

and Associated Companies throughout the world

Visit us on the World Wide Web at:
http://www.pearsoneduc.com

First published 1998

ISBN 0 582 30250 1

British Library Cataloguing-in-Publication Data

A catalogue record for this book is available from the British Library

Transferred to digital print on demand, 2008
Printed and bound by CPI Antony Rowe, Eastbourne

Set by 3 in 11/12 pt Times

DEDICATION

To 'the Accountant'

and to all students on the BSc (Hons) Estate Management
Surveying degree at the University of Glamorgan – past, present
and future.

Contents

Tables

Preface

This book deals with the sources of local authority income which come directly from the taxation of land and buildings. Thus, it explains:
(a) the Uniform Business Rate; and
(b) the Council Tax.
Both these taxes contribute to local authority income and both are based on the value of land and buildings. A brief overview of their development and their role within the context of local government is provided within Chapter 1.

They are very different taxes, with different legislation, procedures and practice. However, there are similarities, although these are relatively few.

For these reasons, the Uniform Business Rate and the Council Tax are dealt with totally separately within this text, which explains the Uniform Business Rate in Chapters 2–11 and the Council Tax in Chapters 13–14. In addition, the criticisms of both taxation systems are also explained (in Chapters 12 and 15 respectively).

This text is written for students on undergraduate courses which lead to qualifications within the profession of property valuation. It therefore focuses on the aspects of the taxes which affect property valuers and managers rather than local authority tax administrators. The text also includes information about how properties are valued for the purposes of local taxation and assumes a knowledge of the so-called traditional methods of valuation which would be usual for level 2 (second-year) undergraduates on such courses.

The book is designed to plug the gap between the available texts, those which are aimed at the practitioners, which are of immense practical use, but do not always explain the basic principles, and those of encyclopedic dimensions which detail everything.

Each chapter is structured in a similar way, beginning with a synopsis of the topic and ending with a check-list. The text includes references to relevant case law and to sections or regulations within the legislation. In this way, readers are able to investigate further primary sources, if required. In addition, a bibliography is provided for further references.

This book is written using 'he', 'him' and 'his' throughout. This is a deliberate policy which, while much against my natural inclination (as any of my students will recognise), does avoid the very awkward 's/he' or (even worse) 'him or her'. It is also the same approach adopted by the learned judges in their decisions, some of which are quoted within the book – so a degree of consistency is maintained. I hope that no one is offended.

The law is stated as it is believed at 1 April 1997 and, therefore, includes the provisions of the Local Government and Rating Act 1997.

Frances Plimmer
April 1997

Acknowledgements

This book has benefited from the valuable contributions made by Roger Collins, David Raley and most especially David Varley. I am extremely grateful to them for their time and expertise.

The presentation of the text has been improved by the careful and conscientious proof-reading of Hilda Sterry-Evans.

The following colleagues have provided support and information which have enabled me to complete the text and I am grateful to Professor Peter Hibberd, Philip Leverton, Jane Daniels, Terry Powell, Ann Cross and Gerwyn Griffiths.

I also acknowledge the kind permission of The Royal Institution of Chartered Surveyors to reproduce the conclusions of the Bayliss Report.

However, without the forbearance and understanding of my family, this book would remain a figment of my imagination and it is to them that I owe the greatest debt.

Abbreviations

Within the text, the following abbreviations are used:

1967 Act	General Rate Act 1967
1988 Act	Local Government Finance Act 1988
1992 Act	Local Government Finance Act 1992
AVD	antecedent valuation date
et seq	and the following
ibid.	as above
ISVA	Incorporated Society of Valuers and Auctioneers
p.	page
pp.	pages
para.	paragraph
reg.	regulation
RICS	The Royal Institution of Chartered Surveyors
RV	rateable value
s.	section
Sch.	Schedule
SI	Statutory Instrument
UBR	Uniform Business Rate
VO	Valuation Officer
VOA	Valuation Office Agency
VT	Valuation Tribunal
YP	Year's Purchase

Appendix G contains a glossary of other terms which may prove useful to the reader.

Acts of Parliament

Cases

Associated Cinema Properties, Ltd. v. Hampstead Borough Council (1943) 1 KB 49 – 2.3.13

B.H.S. Ltd. v. C.B. of Brighton & Burton (VO) (1958) 3 RRC 344 – Appendix E

Barratt v. Gravesend Assessment Committee (1941) 2 KB 107 – 6.5.31; Appendix D

Batty v. Burfoot (1995) RA 299 – 14.3.10

Beath v. Poole (VO) (1973) 228 *EG* 73 – 5.4.32

Bell Property Trust Ltd. v. Assessment Committee for the Borough of Hampstead (1940) 2 KB 543 – 6.4.44

Bradford Metropolitan City Council v. Anderton (1991) RA 45 – 13.3.9

Bradley v. Bayliss (1881) 51 QBD 183 – 2.3.35

Burley (VO) v. A & W Birch Ltd. (1959) 5 RRC 147 – 5.3.17

Burton-upon-Trent (Mayor etc.) v. Burton-upon-Trent Union (1889) 24 QBD 197 – 2.3.40

Burton Latimer Urban District Council v. Weetabix Ltd. and Lee (VO) (1958) 3 RRC 270 – 3.3.26

Byrne v. Parker (VO) (1980) RA 45 – 5.4.8

Calthorpe v. McOscar (1923) KBD 273 – 5.4.26

Cambridge University v. Cambridge Union (1905) 1 Konstam 105 – 6.6.5

Cardiff Corporation v. Williams (VO) (1973) 18 RRC 1 – 6.6.5; 6.6.11

Cardiff Rating Authority & Cardiff Assessment Committee v. Guest Keen Baldwin's Iron and Steel Co. Ltd. (1949) 1 KB 385 – 6.6.5; 6.6.8

Causer v. Thomas (VO) (1957) LT R & IT v. 50/460 – 5.4.25

Chandler (VO) v. East Suffolk County Council (1958) 3 RRC 328: 51 R & IT 411 – 6.6.5

Chesterfield Tube Co. Ltd. v. Thomas (VO) (1970) 3 All ER 733 – 3.2.23; 3.2.40

China Light & Power Co. Ltd v. Commissioner for Rating and Valuation (1996) RA 475

Clement (VO) v. Addis Ltd. (1988) RA 25 – 9.10.3–4; Appendix D

Saunders v. Maltby (VO) (1977) 242 *EG* 299 – 5.4.27–28

Scott (Thomas) & Sons (Bakers) Ltd. v. Davies (VO) (1969) 16 RRC 30 – 6.2.20; 6.7.4

Sheerness Steel Co. plc v. Maudling (VO) (1986) RA 45 – Appendix D

Shell–Mex & B.P. Ltd. v. James (VO) (1960) 8RRC 135; (1961) RVR 106 – 6.6.5

Shrewsbury Schools v. Shrewsbury Borough Council and Plumpton (VO) (1960) 7 RRC 313 – 6.2.18; 6.7.3; 6.7.9

Smith v. Moore (VO) (1972) 17 RCC 377 – 10.13.6

Smith (W.H.) & Son Ltd. v. Clee (VO) (1977) 20 RRC 235; 243 *EG* 677 – Appendix E

Southend-on-Sea (Mayor, etc.) v. White (1900) 65 JP 7 – 2.3.17

Southwark London Borough Council v. Briant Colour Printing Co. Ltd. (1977) RA 101 – 2.3.9

Spear v. Bodmin Union (1880) 43 LT 127 – 3.3.11

Staley v. Castleton (1864) 5 B&S 505 – 5.4.32

Stirk & Sons, Ltd. v. Halifax Assessment Committee (1922) 1 KB 264 – 5.4.4

Stringer (VO) v. J Sainsbury plc (1991) [1992] RA 16 – 7.3.18

Taunton Borough Council v. Sture (VO) (1958) 4 RRC 32 – 6.5.29

Thomas (S.) & Co. (Nottingham) Ltd. v. Emett (VO) (1955) 48 R & IT 761 – 9.4.4

Thomas v. Cross (1951) 44 R & IT 471, 158 *EG* 54 – Appendix D

Thorn EMI Cinemas Ltd. v. Harrison (VO) (1986) RA 125; 279 *EG* 512 – Appendix D

Tomlinson (VO) v. Plymouth Argyle Football Co. Ltd. (1960) 6 RRC 173; 53 R & IT 297; 175 *EG* 1023 – 5.3.19; 6.6.5

Trevail (VO) v. C & A Modes Ltd. (1967) 13 RRC 194 – 6.4.29; 6.4.37; 7.3 19; Appendix E

Trocette Property Co. v. Greater London Council (1974) 28 P & CR 408 – 14.4.6–7

Tulang Properties Ltd. v. Noble (VO) (1984) 272 *EG* 567–8 – Table 7.3

U.D.S. Tailoring Ltd. v. B.L. Holdings Ltd. (1981) 261 *EG* 49 – Appendix E

Union Cold Storage Co. Ltd. v. Phillips (VO) (1973) 236 *EG* 125 – 3.2.39

University of Glasgow v. Assessor for Glasgow (1952) SLT 304 – 3.3.17

Vincent Bach International Ltd. v. Kubbinga (VO) (1994) RA 31 – Table 7.4

Wakefield District Light Railway v. Wakefield Corporation (1908) AC 293 – 2.3.23

Statutory Instruments

The world of rating appears ... to be cloud-cuckoo land, a world of virtual unreality from which real cuckoos are excluded (although it seems that permission to land will be granted to a cuckoo flying in from the real world if it can demonstrate that its presence in cloud-cuckoo land is essential, not merely accidental ...) A valuation for rating purposes must be based on hypothetical, not real, facts.
per Godfrey JA, in *China Light & Power Co. Ltd. v. Comr. for Rating and Valuation* 1995 – 2 HKC 42

Chapter 1

Local taxation

1.1 Synopsis

1.1.1 Local authorities have a variety of sources of income, of which some 35% is derived from the value of land.

1.1.2 Central government levies the Uniform Business Rate (UBR) but gives the revenue raised to local government to spend on services within its area.

1.1.3 Local government levies the Council Tax and uses all the revenue it raises for the benefit of services within its area.

1.1.4 The importance of the taxes based on the value of land to local authorities is significant in terms of revenue raised.

1.1.5 Each tax is considered separately below and in subsequent chapters.

1.2 Local authority functions

1.2.1 Local authorities spend their revenue (not just the revenue raised by the UBR and the Council Tax) in accordance with statutory requirements, and this is illustrated in Table 1.1.

1.2.2 The services which local authorities provide are required by statute. Acting in breach of statute or in the absence of statutory

Table 1.1 Local authority spending 1995–96

	Amount spent (£m)	Proportion of total spent
Education	17,604	40%
Personal Social Services	7,753	18%
Police	6,044	14%
Fire	1,308	3%
Other Home Office	895	2%
Highways and transport	2,213	5%
Libraries, museums and art galleries	775	2%
Parks and recreation	1,168	3%
Planning and economic development	687	1.5%
Waste collection, disposal and regulation	1,054	2%
Other services	4,224	9.5%
Total (error resulting from rounded figures)	44,005	100%

Source: Original figures from *The Chartered Institute of Public Finance and Accountancy, Finance and General Statistics*, CIPFA, 1995, p. 6.

authorisation is *ultra vires* and can be prevented, either by district auditors or by a legal action by ratepayers. Within the limited constraints of their income, the local authorities have the freedom to determine the level of services which they choose to provide.

1.2.3 However, there is only flexibility to determine the level of services if there is flexibility in the amount of revenue received by local authorities.

1.2.4 For 1995–96, the sources of local authority income are indicated in Table 1.2.

1.2.5 Government grants are controlled by central government Originally the grant system was introduced to reduce the level of rates on relatively poor rural areas and to allow those authorities responsible to provide a comparable level of services to that provided within the relatively rich urban areas.

1.2.6 Other sources of income for local authorities include charges for services, which tend to be fixed at or near break-even point and allow no room for flexibility. This is also true of any new borrowing, which is expensive and unpopular. The sale of assets produces little income and reduces capital assets.

1.2.7 Thus, the property-based sources of revenue, i.e. the Uniform Business Rate and the Council Tax, which are buoyant, certain and predictable in yield, are important for local authorities. Though central government fixes the level of the Uniform Business Rate,

Table 1.2 Local authority budget figures 1995–96

	Amount received (£m)	Proportion of total amount received
Central government grants, including the rate support grant	36,426	62.8%
Uniform Business Rate	11,874	20.5%
Reserves	1,056	1.8%
Other income	275	0.5%
Council Tax	8,287	14.3%
Total	57,918	100%

Almost 35% (20.5% and 14.2%) of local authority income is derived from land-based taxes (UBR and the Council Tax).
A total of 83.3% of income is derived from central government, once the Uniform Business Rate is added to other central government sources of funding.
Source: Original figures from *The Chartered Institute of Public Finance and Accountancy, Finance and General Statistics,* CIPFA, 1995, p. 6.

local authorities are free, within central government's capping provisions, to fix whatever level of Council Tax they need to meet their expenditure.

1.2.8   This text considers both taxes from the point of view of the property valuer who is responsible for providing the valuations on which the taxes are levied. Reference should be made to other texts for the details of other aspects of the taxes, such as their administration.

1.3   Uniform Business Rate (UBR)

1.3.1   The present rating system may be known as the National Non Domestic Rate (NNDR) or as the Uniform Business Rate (UBR). This text refers to the Uniform Business Rate (UBR) or 'rate' throughout.

1.3.2   The UBR was introduced into England and Wales with effect from 1 April 1990 alongside the Community Charge (or poll tax) under the Local Government Finance Act 1988 (as amended) as part of the Conservative government's long-standing pledge to reform business rates. The reasons for this are now largely historical, but reference should be made to Appendix A for an overview.

1.3.3   For the purposes of this text, only the current, post-1990, rating system is explained though, where necessary, reference is made to the pre-1990 system, if this is likely to be useful in practice or in analysing critically the existing system.

1.4   Rates defined

1.4.1   The Uniform Business Rate ('rates'), is a tax fixed by central government based on the annual value of non-domestic property, paid by occupiers or, where there are no occupiers, owners, to a local authority which is responsible for providing services within the locality. Local authorities are required to forward the revenue collected to central government, which redistributes it to local authorities for spending.

1.4.2   The UBR is, therefore, an assigned revenue tax, being fixed by and payable to central government and assigned to local government for spending.

1.4.3   There are separate UBRs for England and Wales, and, with effect from 1 April 1995, there is a UBR in Scotland. The UBR for Scotland was set at the same level as that for England (see Table 1.3). This, and other harmonisation measures, will ensure the same rating system throughout Great Britain. (This, of course, excludes Northern Ireland, which currently retains domestic rating.)

Table 1.3 Levels of Uniform Business Rate applicable to date

Year	England	Scotland	Wales
1990–91*	34.8		36.8
1991–92	38.6		40.8
1992–93	40.2		42.5
1993–94	41.6		44.0
1994–95	42.3		44.8
1995–96*	43.2	43.2	39.0
1996–97	44.9	44.9	40.5
1997–98	45.8	45.8	41.5

* Year in which a revaluation came into effect

1.4.4   There are other 'rates', e.g., water rate, sea defence rates, garden rate etc., but these should not be confused with the UBR.

1.4.5   The UBR was fixed by the Secretary of State for the Environment (for England) and by the Secretary of State for Wales (for Wales) for 1990–91 and for 1995–96 (and future years which follow a revaluation), and by the Chancellor of the Exchequer for other financial years (see 1.6).

1.4.6   The rate paid is based on the rateable value of property (see Chapter 5) and is a tax on occupation and, where there is no occupation, on ownership of non-domestic property (see Chapter 2).

1.4.7   *History*

Rates originated in its present form as a nationwide property tax as a result of the Poor Relief Act 1601, the objective of which was to provide a welfare system for

... the necessary relief of the lame, impotent, old, blind and such other among them, being poor and not able to work, and also for the putting of such children to be apprentices ... (s. 1 Poor Relief Act 1601)

1.4.8   The administration was carried out by the Overseers of the Poor and the finance came from the rate levied on:

... every inhabitant, parson, vicar and other, and [on] every occupier of lands, houses, tithes ... coal mines or saleable woodlands in the ... parish. (ibid.)

1.4.9   Later, the role of administration passed from Overseers of the Poor to local authorities, which were authorised to use the income from what was more correctly called the general rate for their many and varied responsibilities.

1.4.10   The UBR has replaced the general rate, which was a tax fixed by local authorities, based on the value of property, paid by local people to the local authority which was responsible for providing services within the locality. The general rate was abolished with effect from 31 March 1990.

1.4.11   With effect from 1 April 1990, the responsibility of local authorities to fix the rate was transferred to central government, and the liability of occupiers of domestic property to pay rates was abolished and transferred to the Community Charge (or poll tax), which was replaced by the Council Tax in 1993 (see Chapters 13–15 and Appendix A).

1.4.12   A more detailed historical timetable is contained in Appendix B.

1.5   Ratepayers

1.5.1   Rates are paid by occupiers of non-domestic property and by owners where there is no occupier of non-domestic property.

1.5.2   This occupiers' liability stems from the original 1601 requirement for rates to be paid by:

... every inhabitant, parson, vicar and other, and ... every occupier of lands, houses, tithes ... coal mines or saleable woodlands in the ... parish (Poor Relief Act 1601).

1.5.3   By 1966, 'inhabitants' etc. had been exempted from liability to rates, leaving only the 'occupier' responsible for the tax.

1.5.4   Also in 1966, central government extended rate liability to owners of empty property. Empty properties continue to receive the benefit of such public services as street light and cleansing, police and fire services, so it was considered equitable that their owners should contribute towards the provision of such services.

1.5.5   Historically, therefore, the liability to rates is an occupiers' liability, which is specifically defined as that of an occupier in 'rateable occupation' (see Chapter 2) for rating law. With effect from 1 April 1990, there is a mandatory, statutory liability for owners to pay rates on their empty non-domestic property.

1.6   Fixing of the rate

1.6.1   The level of the UBR is fixed for each year based on the level of UBR in the previous year plus inflation, which is measured according to the rate of inflation for the twelve months up to September of the year before that for which the UBR is fixed (i.e., for the 1996–97 UBR, the rate of inflation used was that for the twelve months beginning September 1995) (Sch. 7, para. 3 Local Government Finance Act 1988) as illustrated in Figure 1.1.

1.6.2   This is, in fact, the maximum by which the UBR can be increased and there is the opportunity to fix the UBR at a level lower than that achieved by the formula. However, the formula ensures that the UBR can increase only in line with inflation, and levels of UBR to date have, in fact, risen only in line with the (September to September) level of annual inflation. The levels of UBR applicable to date are listed in Table 1.3.

1.6.3   There is one UBR set for England, one for Wales, one for the City of London, another for the Isles of Scilly and, with effect from 1 April 1995, one for Scotland. For 1997–98, the Secretary of State for the Environment set a UBR of 45.8p for England. The same level of UBR was set for Scotland and the Secretary of State for Wales set a UBR of 41.5p for Wales.

$$\frac{A \times B}{C}$$

where   A = last year's UBR

B = last year's RPI for September

C = previous year's RPI for September

For example: for 1997–98, the UBR for England is calculated, as follows:

$$\frac{A \times B}{C}$$

where   A = UBR 1996–97

B = RPI for September 1996

C = RPI for September 1995

$$\frac{44.9 \times 153.8}{150.6} = 45.8p$$

Figure 1.1   Amount by which the Uniform Business Rate is increased every year (except for the year following a revaluation). (Schedule 7 para. 3 Local Government Finance Act 1988.)

1.6.4   In the year following a revaluation of all rateable property, in which new rating lists come into force, the UBR is determined by finding that figure which, when multiplied by the total of all the rateable values in the country, raises the same amount of rate money as was raised from rateable property in the previous year (Sch. 7 para. 4, 1988 Act). The figure so determined is index-linked by an amount which does not exceed the rate of inflation (see Figure 1.2).

1.6.5   This dependence on a formula to establish the level of rate liability enables non-domestic ratepayers to plan and budget for their future rate liability without relying on the politically-motivated spending policies of their local authorities, which, prior to 1990, were able to set their own level of rate, with only limited central government interference. It also means that, once the September RPI figures are released, the level of UBR for the forthcoming financial year can be predicted with accuracy.

1.6.6   The City of London has been treated differently for UBR

$$\frac{A \times B \times D}{C \times E}$$

where   A = last year's UBR

B = last year's RPI for September

C = previous year's RPI for September

D = total rateable value of all the properties
in the 'old' list and

E = total rateable value of all the properties
in the 'new' list.

Figure 1.2   Amount by which the Uniform Business Rate is
increased for the year following a revaluation. (Schedule 7 para. 4
Local Government Finance Act 1988.)

purposes. Because of its very low resident population and the high
costs of providing services to support the massive influx of day-
time commuters, the City of London is able to raise a separate rate.
This was originally introduced in order to avoid a Community
Charge of about £8,450 per adult, compared with a predicted
national average of £280.

1.7   Rate collection

1.7.1   Rates is a central government tax raised specifically to support the
provision of local authority services. The income is, however,
collected and spent by local authorities.

1.7.2   Having collected the tax, the level of which is set by central
government (see 1.6 above), the revenue is, notionally, pooled and
redistributed between all local authority areas based on the
number of adults in the area. There is one pool for England and
one for Wales.

1.7.3   The amount paid into the pool for England from 1990–91 to
1997–98 is estimated to total £93.5 billion (thousand million) and
equals the amount distributed to local authorities over the same
period.

1.7.4   In practice, local authorities keep the revenue they raise and, if
they collect more than they are entitled to under the scheme for

distribution, pay the excess to central government, or, if they collect less than they are entitled to under the scheme for distribution, receive the balance from central government.

1.7.5   Should a local authority, for whatever reason, fail to collect the UBR as established by central government, it is credited with the amount due and no replacement funds are available from central government. Local authorities have no alternative but to make up the deficit from Council Tax revenue. This is not designed to be popular!

1.8   Amount of rates paid

1.8.1   The amount of rates paid is arrived at by multiplying the UBR by a value for the property which is called the rateable value (RV). Thus: rates paid = UBR × RV.

1.8.2   The UBR or rate is fixed by central government annually for the financial year (see 1.6 above), and is always cited in terms of a penny rate, e.g., 41.5p (£0.415).

1.8.3   The rateable value is a net annual rental value of the property (see Chapter 5) and is the value to which the rate is applied in order to calculate the amount of rates payable (rates paid = UBR × RV).

1.8.4   Thus, any one who knows the rateable value of the property occupied and the UBR can easily calculate the annual liability. For example: (UBR) 41.4p × (RV) £50,000 produces rates paid of £20,700 per annum.

1.8.5   Although the liability to pay rates is a daily liability, the level of the rate (UBR) is fixed for a full financial year. Rates can, however, be payable in ten equal monthly instalments (see Chapter 11).

1.9   Council Tax

1.9.1   The Council Tax was introduced following the abolition of the Community Charge (poll tax) on 1 April 1993 (see Appendix A).

1.9.2   The Council Tax is paid by occupiers and owners of dwellings and comprises two elements: the personal element represents 50% of the tax payable; and the property element the other 50% of the tax (see Chapter 13).

1.9.3   The level of Council Tax is fixed annually by each local (billing) authority.

1.9.4   Other authorities, such as police authorities, request a share of the revenue from the billing authority. This is called precepting. Indeed, in Wales, police authorities are the only major precepting authorities following the 1996 local government reorganisation.

1.10   Conclusion

1.10.1   As explained above (1.8), rates paid = UBR × rateable value (RV), with the rateable value being fixed by property valuers for each financial year. Similarly, the level of Council Tax is, in part, dependent upon the value of the dwelling in which the taxpayer lives.

1.10.2   Property valuers are, therefore, involved in fixing both the rateable value of non domestic property and the value of domestic property on which the liability to Council Tax is based. The rateable value and the value for Council Tax purposes are strictly defined (see Chapter 5 and Chapter 14, respectively), and statute and case law are used as well as basic principles of valuation to establish those values.

1.10.3   All traditional methods of valuation are used to produce rateable values within the legal framework laid down by rating statutes and judicial decisions (see Chapter 6).

1.10.4   Having fixed a rateable value or the value for Council Tax purposes, it may be necessary to challenge or defend that value (see Chapters 9, 10 and 14). Again, valuers are involved in the process, even if the challenge is on legal grounds, since the effect of the law is likely to be a change in the value of the property.

1.10.5   However, knowledge of traditional valuation techniques and the values within the locality are not sufficient to assess such values. A valuer needs to understand rating law, the definitions of 'value' in both the rating and the Council Tax provisions, and to apply valuation skills within that law.

1.10.6   The law is complex but, despite the changes introduced in 1990, relatively well established. Many of the principles result from judicial decisions, which makes for a vast amount of complex legislation. The 1990 Uniform Business Rate was introduced by the Local Government and Finance Act 1988 and the detail provided by a large number of subsequent statutory instruments and, together, these have given a definitive legislative basis of rating valuation, but this does not alter the fact that the study of rating law can be confusing and can only be truly understood by working within the system.

1.10.7   The aim of this text is:
(a)   to explain the law and to include references to more detailed sources for those who wish to study in further depth;
(b)   to facilitate the solution of practical problems, including the carrying out of valuations; and
(c)   to encourage informed debate about the reform or the replacement of the current system of local property taxation, both as a source of local authority revenue and as a basis for taxing property.

1.10.8   However, despite the almost inevitable concentration of this text on the law of rating which is directed towards fixing a rateable value for property, it is important to remember that land taxation is only a means to an end, i.e., the raising of revenue by the taxing of land and buildings.

1.10.9   Despite its recent reform, the Council Tax has problems (see Chapter 15), which are increasing in severity as they remain unattended.

1.10.10   Similarly, the existing rating is not a perfect system and continues to be the subject of debate (e.g. RICS, 1996, *The National Committee on Rating. Improving the System,* The Royal Institution of Chartered Surveyors (The Bayliss Report), see Appendix C).

1.10.11   It is important that those professionals who operate the local property taxation systems, which raise over £20 billion in the UK every year, should be part of this informed debate.

Chapter 2

Liability to pay rates

2.1   Synopsis

2.1.1   Since 1990, rates are paid both by occupiers of rateable property (specifically, a hereditament) and by owners of unoccupied rateable property (or hereditaments).

2.1.2   However, in each case, the liability to rates is a daily liability. (This stipulation was originally introduced to parallel the Community Charge, which was also a tax levied on a daily basis.)

2.1.3   Liability to occupied rates exists for rateable occupiers, i.e., those who are in actual possession, exclusive occupation, beneficial occupation and permanent occupation of their rateable property.

2.1.4   Liability to unoccupied rates exists for owners of empty property, who pay only half of the level of rates paid by occupiers.

2.2   Occupied rate liability

2.2.1   Before rates are payable, there must be, on any chargeable day of the chargeable financial year:
(a)   a person (ratepayer) in occupation of all or part of the property which is rated (called a hereditament); and
(b)   rateable property (specifically, a hereditament – see Chapter 3) shown in a local non-domestic rating list (s. 43 (1) 1988 Act) (see Chapter 8).

2.2.2   A 'chargeable day' is a day on which both (a) and (b) above occur (s. 43 (1) 1988 Act).

2.2.3   A person is liable to pay a central rate to the Secretary of State if, on any day, that person's name is shown on the central rating list (s. 54 (1) 1988 Act and see Chapter 8).

2.2.4   In addition to the above, an occupier must be in rateable occupation of the property and this is defined in case law.

2.3   Rateable occupation

2.3.1   In determining whether a hereditament (defined in Chapter 3) is 'occupied' and who is the 'occupier', section 65 (2) of the Local Government Finance Act 1988 applies the same rules as those applied under earlier legislation. Thus, the four tests of rateable occupation set out in *John Laing & Son Ltd. v. Kingswood Area Assessment Committee,* (1949), together with most of the judicial decisions on the circumstances in which rateable occupation occurs, continue to apply.

2.3.2   *Occupation*
The liability to pay rates was, originally under the Poor Relief Act of 1601, and remains, an occupier's liability.

The occupier, not the land, is rateable; but the occupier is rateable in respect of the land which he occupies. (Lord Russell of Killowen in *Westminster City Council v. Southern Rail Co.* (1936) at p. 529.)

2.3.3   Although it is often said that a particular property is rateable, this is incorrect. What is legally correct is that an occupier is rateable in respect of his occupation of the land he occupies. On this basis, if there is no occupier, there is no liability to the occupied rate (although there is an unoccupied rate, payable by the owner – see 2.4).

2.3.4   In 1949, the Court of Appeal considered the case of *John Laing & Son Ltd. v. Kingswood Assessment Area Assessment Committee*, in which it confirmed actual possession, exclusive occupation, beneficial occupation, and occupation of sufficient permanence as essential elements of rateable occupation. The absence of any one of these will cause the occupation in question to be not rateable. These elements are considered in greater detail below.

2.3.5   *Actual possession*
Actual possession is hard to define. It refers to 'possession' in the sense of the presence of the occupier on the land but does not include 'possession' in the sense of an owner's legal title to land.

2.3.6 In *R. v. St. Pancras Assessment Committee* (1877) Lush J, said (at p. 588):

> Occupation includes possession as its primary element, but it also includes something more. Legal possession does not of itself constitute occupation. The owner of a vacant house is in possession, and may maintain trespass against anyone who invades it, but as long as he leaves it vacant, he is not rateable for it as an occupier. If, however, he furnishes it, and keeps it ready for habitation whenever he pleases to go to it, he is an occupier, though he may not reside in it one day in a year.
>
> On the other hand, a person who, without having any title, takes actual possession of a house or a piece of land, whether by leave of the owner or against his will, is the occupier of it.

2.3.7 Thus, the courts made a clear distinction between legal possession and actual possession.

2.3.8 Legal possession means legal title to use land and is very much a matter of law. Legal possession alone will not create liability to pay rates (*Westminster City Council v. Southern Rail Co.* (1936)).

2.3.9 Actual possession means the physical presence of the occupier on the land or the acts by which he makes use of or controls the land (*Southwark London Borough Council v. Bryant Colour Printing Co. Ltd.* (1977)). The existence of actual possession is very much a matter of fact, and in order for someone to be rated as an occupier, he must be in actual possession.

2.3.10 It follows from this that if liability to pay rates is based on the fact of actual possession, the legal title of the 'occupier' becomes irrelevant in determining that liability. So, a trespasser can be a rateable occupier and, therefore, liable to rates, provided he has actual possession.

2.3.11 Where an owner has any use at all of land, he is the rateable occupier of it, unless it can be shown that someone else actually occupies it (*Westminster City Council v. Southern Rail Co.* (1936)). (This should be distinguished from the separate obligation on owners of unoccupied property to pay rates, which is considered in 2.4.).

2.3.12 Thus, a distinction is made between legal possession, which does not of itself create liability to occupied rates, and actual possession, which must exist before liability to pay occupied rates can arise.

2.3.13   A mere intention to occupy property is insufficient to create rate-
able occupation. Intention, plus something more, must be present
*(Associated Cinema Properties, Ltd. v. Hampstead Borough
Council* (1943)). Thus, there must be an act of user, which, taken
together with intention, will create rateable occupation *(R. v.
Melladew* (1907)).

2.3.14   In *R. v. Melladew* (1907), the Court of Appeal recognised that the
form of occupation, which is the basis of liability, necessarily
varies with the nature of the property involved and that the test
must be whether the person to be rated has such use of property as
the nature of the property and of the business connected with it
renders it reasonable to infer was anticipated when the property
was acquired or retained.

2.3.15   Although land may be used even though no positive acts are
carried out with regard to it, for occupation to arise, something
must actually be done on the land – either on the whole or on a part
in respect of the whole.

2.3.16   Occupation can result even from a slight use of land (*Liverpool
Corporation v. Chorley Union and Withnell Overseers* (1913)). In
the days before 1929, when agricultural land was rated, it was
recognised that land used for growing crops, etc., was rateable,
even though no-one set foot on it for months, or even years. In
Newcastle City Council v. Royal Newcastle Hospital (1959), Lord
Denning said (at p. 255):

> ... everyone would say that a farmer 'occupies' the whole of his farm
> even though he does not set foot on the woodlands within it from one
> year's end to another.

2.3.17   Because rates are based on an annual tenancy (see Chapter 5),
cases arise where, because trade is seasonal during the winter
months, premises are closed for part of a year, and only tenant's
fixtures and fittings and some chattels remain. In such cases, the
tenants are rateable as the occupiers during the winter months
(*Southend-on-Sea (Mayor, etc.) v. White* (1900) and *Gage v. Wren*
(1903).

2.3.18   This is not as unreasonable as it may at first appear. Rates paid are
based on the annual rental value of a property, determined according to a statutory definition (see 5.3.1). If a property is only
profitable for part of the year, then the rent paid is based on that
fact and, presumably, is lower than if profit lasted for twelve
months. If rates are based on this lower rent, then the seasonal

profit of the occupation has been taken into account, and to make an allowance in the value on which rates are levied for the short season would be to compensate the ratepayer twice.

2.3.19 Certain acts do not give rise to occupation. For example, buildings in the course of construction or alteration are not rateable because there is no occupier. Parts of a building site may, however, be used so as to be rateable in the occupation of the builders. For example, builders' huts, when occupied in one location for a sufficient length of time, have been held to be rateable (see 2.3.49).

2.3.20 When structural alterations are made to a building which continues to be rateable, then the works may constitute grounds for a reduction in the valuation (see also 2.4.42).

2.3.21 Similarly, acts attributed solely to ownership do not render the owner rateable – the occupier is rateable in respect of the property he occupies. If there is no occupier, then there can be no liability to occupied rates (but see Unoccupied rate liability at 2.4). Acts of an owner which do not give rise to occupation include the preservation of an empty house from damage or deterioration.

2.3.22 Thus, to summarise, one of the essential elements of rateable occupation is actual possession. For actual possession to be established, the property must be used, no matter how trivial the use: intention alone is insufficient.

2.3.23 *Exclusive occupation*
Exclusive occupation involves the use of land by someone who, by title or by the nature of the occupation, can exclude everyone else from using the land in the same way (*Wakefield District Light Railway v. Wakefield Corporation* (1908)). Such a person may be a rateable occupier even if someone else can use the land in the same way, provided the rateable occupier's occupation remains paramount (see 2.3.30).

2.3.24 There are occasions when separate occupations will exist in respect of the same land. It may be that, because of the exclusive nature of each of the uses, separate occupations exist in a situation where two (or more) occupiers are able to use the same land but in different ways and each is able to exclude the other(s) from using the land in the same way they do (*Holywell Union v. Halkyn District Mines Drainage Co.* (1895); *Ryan Industrial Fuels Ltd. v. Morgan (VO)* (1965); *Pimlico Tramway Co. v. Greenwich Union* (1873)).

2.3.25   Although title is irrelevant in determining rateable occupation, it may be considered where the facts are ambiguous. In such cases, there is a presumption that the owner of land is the occupier, until it is shown that occupation is in the hands of another.

2.3.26   The presumption was used, for example, in *Liverpool Corporation v. Chorley Union Assessment Committee and Withnell Overseers* (1912), in the case of a slight user in which reference to title was necessary to establish occupation.

2.3.27   Someone without title to the exclusive occupation of land may be the rateable occupier if his occupation is exclusive in fact. If the character of the use he makes of the land is such that others are excluded from using it in the same way, his occupation will (if the other ingredients of rateable occupation are satisfied) be rateable.

2.3.28   In determining whether an occupation is exclusive in fact, it is necessary to consider whether the person has the enjoyment of the premises 'to the substantial exclusion of all other persons' (Lord Russell of Killowen in *Westminster City Council v. Southern Rail Co.* (1936) at p. 532).

2.3.29   There is a substantial number of cases where exclusive occupation, and therefore rateability, have been held not to exist. Rights of way, for example, will often fail to be exclusive.

2.3.30   It sometimes happens that there are two persons involved with the use of land for the same purpose and the question arises which of the two is the rateable occupier. The occupation which is rateable is the 'paramount occupation'.

2.3.31   In *Westminster City Council v. Southern Rail Co.* (1936), the principles to be applied in determining paramount occupation were laid down by Lord Russell of Killowen as follows:

> ... if the owner of the hereditament (being also in occupation ...) retains to himself general control over the occupied parts, the owner will be treated as being in rateable occupation; if he retains to himself no control, the occupiers of the various parts will be treated as in rateable occupation of those parts. (ibid. at p. 530)

2.3.32   From the above, the general principle is that if an occupying owner retains control over the use of parts occupied by another or others, the owner is in rateable occupation of the whole and is, therefore, liable to pay rates. Where the occupation of parts of a hereditament is by two parties, it is the person having paramount control, or whose occupation is paramount, who is the rateable occupier of the jointly-occupied parts.

2.3.33   In applying these general principles, certain rules have been established:

(a)   for the control of the landlord to be paramount, his control must be over the use made of the shared premises;

(b)   the amount of control reserved must be examined to see to what extent its exercise would interfere with the enjoyment by the occupier of the premises in his possession for the purposes for which he occupies them, or would be inconsistent with his enjoyment of the premises to the substantial exclusion of all other persons;

(c)   what is important is the nature of the occupation, not the legal title from which it is derived. It is immaterial whether the title to occupy is attributable to a lease, a licence or an easement;

(d)   control over access to the occupied premises without control over their use will not create occupation;

(e)   the imposition of regulations and/or byelaws over the use of premises will not prevent the rateability of the occupier on whom these are imposed if it can be shown that they are in the nature of restrictive covenants;

(f)   the non-rateability of a lodger does not depend upon the continued presence of the landlord, but upon the fact that the landlord retains control over the use of the whole property;

(g)   what is important is not the terms of the grant, but the actual occupation (which may be greater or less than the grant).

2.3.34   From case law which was decided at a time when domestic property was rated, it was accepted that a 'lodger' (in the popular sense) was not rateable, while the tenant of a flat (generally speaking) was rateable. A lodger's room consisted of part of a larger building, but, in most cases, the lodger occupied subject to the general control which the landlord exercised over the whole property, including the rooms let to the lodger. The occupation by a tenant of a 'flat' was usually rateable because, although he occupied part of a larger building, the tenant was usually free from the control of a landlord over the use of that flat. These principles have been applied to occupations of commercial premises.

2.3.35   The distinction between a lodger and an occupier was made in *Bradley v. Bayliss* (1881), a case concerning parliamentary franchise, but which is referred to in rating law:

... where the owner of a house does not let the whole of it, but retains a part for his own use, and resides there, and does not let out the passages and staircases to the outer door, but only lets to the tenants the right of ingress and egress; and where the owner retains the control over ... where he retains the right to interfere, and to turn out trespassers, and the like; there I consider the landlord is the occupying tenant of the house, and the inmate ... is a lodger. That is one extreme case.

... where the landlord lets out the whole of the house in separate apartments, ... so as to demise the passages, reserving simply to each inmate of the upper floors the right of ingress and egress over the lower passages, parting altogether with the whole legal ownership, and retaining no control over the house, there ... the inmates are occupying tenants and are rateable as such. That is another extreme case.

There are a great number of intermediate cases which I can deal with only as they arise. (per Jessel MR at p. 195)

2.3.36 *Beneficial occupation*
The third requirement of rateable occupation is beneficial occupation, which means that the occupation should be of value to the occupier, for which a tenant would give a rent which is greater than the outgoings on the property. It does not mean that the occupation should be profitable.

2.3.37   In the case of private property used to make a profit, it has never been doubted that, if a profit is made which is totally absorbed by the agreed rent to the landlord, the tenant is in beneficial occupation and, therefore, rateable.

2.3.38   However, problems arose regarding the occupation of premises which were not used to make a profit and case law established that such occupiers as charitable organisations which make no profit may still be in beneficial occupation of their premises.

2.3.39   The test in rateability is not whether the property will produce a profit, but whether it will produce a rent.

2.3.40   The occupation of such properties as schools, where no profit is allowed to be made, and sewerage systems, which cannot be occupied at a profit, may still be beneficial, since someone would pay a rent for the premises, particularly if occupation of such premises was required to fulfil a statutory duty (*West Bromwich School Board v. West Bromwich Overseers* (1884); *Burton-upon-Trent (Mayor etc.) v. Burton-upon-Trent Union* (1889)).

... the appellants would not be able to carry out their statutory duties as to the disposal of sewage at any other place at a smaller expense. If the land and pumping-station in question belonged to a private owner, he would let, and the appellants would hire them at a yearly rent sufficiently high to support the present rate. (ibid. at p. 198)

2.3.41   Land for which no rent would be paid is said to be 'struck with sterility'. Land dedicated to the public as a highway is 'struck with sterility', because, while the dedication lasts and the public has such extensive rights of user over it, no tenant would pay rent for it. (The 'public' is not a rateable occupier.)

2.3.42   Public parks and recreation grounds subject to extensive public rights of user may not be in the beneficial occupation of the local authority which owns and maintains them. The test is whether the public has 'free and unrestricted use' of the park, to the extent that no tenant would pay a rent for it. (*Lambeth Overseers v. London County Council* (1897) (the Brockwell Park Case)). Public parks are now exempt rates (see 4.3.29), but the principles are applied to other similar occupations.

2.3.43   Provided there is actual occupation (i.e., acts of user), and the occupation is exclusive, the fact that the hereditament may not be used to the full by the occupier is irrelevant. Rateability can only be avoided by abandoning the use of the hereditament altogether.

2.3.44   There are certain acts which do not constitute beneficial occupation. An owner is not rateable just because he keeps a caretaker on the premises, where the caretaker's duties are limited to the preservation and protection of the premises and facilitating the entry of the owner *(Yates v. Chorlton-upon-Medlock Union (1833))*.

2.3.45   Although the keeping of furniture or other chattels in a property not otherwise used generally amounts to beneficial occupation, where furniture or goods are left behind when premises are vacated because it is simply not worth removing them, this does not amount to beneficial occupation.

2.3.46   Briefly then, beneficial occupation means occupation which is of value, i.e., where the property is capable of supporting a rent or where, because of the nature of the occupier, a rent would be paid to occupy that property.

2.3.47   *Permanent occupation*
To be rateable, occupation must have a sufficient degree of permanence, and the courts will consider not only the length of the occupation but its character as well. In *R. v. St. Pancras*

Assessment Committee (1877), the question of permanence was discussed by Lush J at p. 589:

> As the poor-rate is not made day by day, or week by week, but for months in advance, it would be absurd to hold that a person, who comes into the parish with the intention to remain there a few days or a week only, incurs a liability to maintain the poor for the next six months. Thus a transient, temporary holding of land is not enough to make the holding rateable.

2.3.48   Although the liability to the Uniform Business Rate under the 1988 Act is a daily liability, it seems that the principle of a sufficient degree of permanence is still applied to the nature of the occupation.

2.3.49   Builders' huts in position for more than a year have been held to be rateable, and in *London County Council v. Wilkins (VO)*, (1956), it was said at pp. 48–9:

> An occupation is not the less permanent because it is that of a lessee who holds under a lease for a fixed term ... an occupation can be permanent even though the structure or other chattel which is the means of occupation is removable on notice.... It may be that permanent signifies no more than continuous, as opposed to intermittent, physical possession of the soil ... a tenant at will has an occupation that is sufficiently permanent.... The rate is an annual impost on the occupier in respect of his profitable occupation of land. ... If such an occupation in fact endures for a year or more, I do not see why the occupier should not contribute to the current fund of the rating area for that period.

2.3.50   Liability to have an occupation terminated by notice or at will does not prevent the occupation from being rateable. Thus, a weekly tenant of a property has a sufficient degree of permanence for rateability, because occupation in fact continues for longer.

2.3.51   In the case of markets where stall-holders erect their stalls on a regular weekly basis, rateability is not avoided because the market is only held on certain days. It is the occupier of the market as a whole, rather than individual stall-holders, who would be rateable in such circumstances.

2.3.52   The character of the occupation as well as length is also important. Short-term quarrying of nine months and even six months has been held sufficiently permanent, because of the nature of the occupation, and a 'twelve-month working rule' has been rejected by the courts.

2.3.53   Thus, the degree of permanence needed for rateable occupation depends on the actual length of the occupation, as well as the nature of the occupation. The rate is fixed for twelve months, so liability to pay rates results from a relative permanence of occupation.

2.3.54   It is recognised that all four essentials of rateable occupation must be present before liability to pay occupied rates exists.

2.4   Unoccupied rate liability

2.4.1   In addition to the occupier's liability to pay rates, there is a liability on owners to pay the Uniform Business Rate which is levied on their unoccupied non-domestic rateable property (see 1.5.4 and Chapter 3).

2.4.2   Owners often pay rates in their capacity as occupiers, and although the rating of owners (in their capacity of owners) existed prior to 1990, the provisions laid down in the Local Government Finance Act 1988 mark a radical departure from the earlier situation.

2.4.3   *Charging provisions*
The rating of owners of unoccupied hereditaments (rateable property) is compulsory, although they are required to pay only half the normal (occupied) level of rates.

2.4.4   Before the unoccupied rate becomes payable, the following conditions (s. 45 (1) 1988 Act) must be fulfilled on any day in the year, namely:
(a)   a relevant non-domestic hereditament (defined in Chapter 3) must be shown in a local non-domestic rating list (see Chapter 8);
(b)   all of that hereditament must be unoccupied;
(c)   the hereditament must fall within a description prescribed by the Secretary of State for the Environment; and
(d)   the ratepayer must be the owner of the whole of the hereditament.

2.4.5   The 'owner' of a rateable property is 'the person entitled to possession of it' (s. 65 (1) 1988 Act). It is not necessary to own a freehold or leasehold interest in the land in order to be an 'owner' for rating purposes.

2.4.6   Despite the origins of rates as a tax on the occupation of land, the rating of owners has occurred because it is generally accepted that owners of property which benefit from services provided by the local authority should contribute towards this provision, and rating law has been altered to reflect this change of attitude (although the current provisions have been criticised, see, for example, Appendix C, paras. 67–9).

2.4.7   It is also true that making owners liable for rates in some situations facilitates rate collection for the billing authority. By making the owner responsible for rate payment, either by altering the hereditament from several to one, or by making the owner a 'go-between', rate collection becomes easier and more certain and, therefore, cheaper for all ratepayers.

2.4.8   However, where owners are liable for the unoccupied rates, they pay in respect of properties which are empty.

2.4.9   *Chargeable amount*
The unoccupied rate due from the owner/ratepayer is half that payable by an occupier, i.e., it is one half of the chargeable amount for each chargeable day (on the basis of the circumstances which existed immediately before the day ends).

2.4.10   The chargeable amount is calculated by multiplying the rateable value (see 5.3.1) shown in the rating list (see Chapter 8) by the Uniform Business Rate and dividing the result, first by the number of days in the chargeable financial year and then by two, so that only half the normal rate liability is payable by owners of unoccupied property (s. 45 (4) 1988 Act).

2.4.11   Where the owner/ratepayer is a charity or trustee for a charity, and it appears that when next in use the hereditament will be wholly or mainly used for charitable purposes, the amount payable is one-fifth of the unoccupied rate, as calculated above (s. 45 (5) and (6) 1988 Act) (see 4.5.1–2).

2.4.12   *Empty properties*
The provisions relating to the rating of unoccupied property are laid down in s. 45 of the 1988 Act. Particular classes of un-occupied property to which these provisions apply are specified by the Secretary of State.

2.4.13   Liability to be rated in respect of unoccupied property attaches to owners (i.e., the person entitled to possession) of the relevant hereditaments.

2.4.14   The hereditament must have been unoccupied for a continuous period exceeding three months (s. 2 (a) Non-Domestic Rating (Unoccupied Property) Regulations 1989 (SI 1989 No. 2261)). Where the hereditament has been unoccupied and then becomes occupied for a period of less than six weeks, that period of less than six weeks' occupation is considered as a period of unoccupation, for the purposes of calculating the three-month period. Thus, it is not possible to extend the three-month 'rate-free' or 'void' period by installing very short-term tenants (ibid. s. 3).

2.4.15   Special provisions exist (s. 44A 1988 Act) to require the valuation officer (who is responsible for valuing the property – see 5.2.4–6) to apportion the assessment of property, part of which is temporarily empty or which is being occupied and vacated in stages, at the request of the billing authority. The apportioned assessment enables the billing authority to levy rates solely on the occupied area of the property.

2.4.16   *Exemptions*
Hereditaments which are exempt from the unoccupied rate are listed in the Non Domestic Rating (Unoccupied Property) Regulations 1989 (SI 1989 No. 2261), as amended by the Non Domestic Rating (Unoccupied Property) (Amendment) Regulations 1995 (SI 1995 No. 549).

2.4.17   Thus, an owner of an unoccupied hereditament is not liable to pay rates on property which is empty where:
(a)   the owner is prohibited by law from occupying or allowing the hereditament to be occupied;
(b)   the hereditament is kept vacant because either the Crown or any local or public authority intends prohibiting occupation, or acquiring it;
(c)   the hereditament is the subject of a building preservation notice (ss. 54 and 58 Town & Country Planning Act 1971);
(d)   the hereditament is included in the Schedule of monuments compiled under the Ancient Monuments and Archaeological Areas Act 1979;
(e)   the empty property is an industrial hereditament (see 4.3.29);
(f)   the empty property is a warehouse (see 4.3.29);
(g)   the empty hereditament has a rateable value not exceeding £1,500 (with effect from 1 April 1995).

2.4.18   There are other exemptions which basically refer to the status of the owner, e.g., if he is entitled to possession in his capacity as a personal representative of a deceased person, as a trustee under a deed of arrangement, or as a liquidator under the Insolvency Act 1986.

2.4.19   The billing authority has the discretionary power to reduce or remit the payment of rates on unoccupied property if it is considered that enforcing payment will cause hardship to the person liable (s. 49 1988 Act – see 4.5.8–9).

2.4.20   *'Mothballed' factories*
Despite the principle of rating law that any use of property constitutes beneficial occupation and, therefore, rateable occupation (see 2.3.36–46), plant and equipment located in an industrial property which is otherwise unoccupied will not create a liability for unoccupied rates (s. 65 (5) 1988 Act).

Thus, where for reasons of an industrial recession, for example, a factory is fully equipped but not otherwise in occupation, no liability for unoccupied rates exists.

2.4.21   *Newly-completed or structurally-altered buildings*
When a property becomes empty, either following its original construction or following substantial structural alterations, there is a three-month 'rate-free' or 'void' period before rates become payable.

2.4.22   However, while such a building is undergoing building works, the owner is usually seeking a tenant, unless the property is purpose-built for a particular occupier who will pay the occupied rate liability on occupation.

2.4.23   Should it become likely that an occupier will not be available for the building once it is completed, the owner is faced with the prospect of having an empty building, producing no income, for which he is liable to pay unoccupied rates.

2.4.24   In such a case, an owner may be tempted to delay completion of the building until a potential tenant is secured, in order to avoid the unoccupied rate liability.

2.4.25   If the building is incomplete to the extent that no one will pay a rent for it, then it is incapable of being beneficially occupied (see 2.3.36–46) and therefore incapable of rateable occupation. If no rent would be paid for it, no rateable value can be fixed.

2.4.26 If the building, while still incomplete, is capable of beneficial occupation as it stands, then the assessment is likely to be substantially below that which the finished building would command and, for that reason, the rates bill will be lower.

2.4.27 However, the billing authority is able to have the building treated as complete under certain circumstances and, as a result, unoccupied rates will be paid by the owner on the full value, assuming the building to be complete.

2.4.28 ***Completion notice***
With effect from 1 January 1990, a billing authority has discretionary power to serve a completion notice on the owner of a non-domestic building where the billing authority is of the opinion that the building has been completed, or that the work remaining to be done is such that it can reasonably be expected to be completed within three months (s. 46A and Sch. 4A, 1988 Act, as amended).

2.4.29 If, in the opinion of the billing authority, the building is complete or the work outstanding can be completed within three months, it can serve a completion notice on the owner, stating that, from a date specified within the notice, the building will be treated as complete. The three-month 'void' period will run from that date, after which the unoccupied rate will become payable (Sch. 4A, para. 1 (1) 1988 Act).

2.4.30 Where a building is considered to be complete, the date specified is the date on which the completion notice is served (Sch. 4A, para. 2 (3)).

2.4.31 'Completed' means ready for occupation for the purpose for which it was constructed or adapted (*Ravenseft Properties Ltd. v. Newham London Borough Council* (1975)).

2.4.32 For the purposes of Sch. 4A, the owner is 'the person entitled to possession of the building' (Sch. 4A, para 10 (2) 1988 Act).

2.4.33 The valuation officer may prevent the service of a completion notice by writing to the billing authority (Sch. 4A, para. 1 (1) 1988 Act).

2.4.34 A completion notice must specify the building to which it relates and state the day which the authority proposes as the completion day for the building (Sch. 4A, para. 2 (1)).

2.4.35 The completion notice will take effect even if the work outstanding is not carried out within the time-limit specified in the Notice. What is relevant is that, in the opinion of the billing authority, the work outstanding can be completed within the time-limit set.

2.4.36 The effect of the completion notice is that from the date specified (or such other date as is agreed or determined by the valuation tribunal) the building is to be treated as unoccupied and, three months after that date, the owner becomes liable to the unoccupied rate (Sch. 4A, para. 5 1988 Act).

2.4.37 Appeal against a completion notice is to the valuation tribunal (see 9.2) (Sch. 4A, para. 4 (1)).

2.4.38 If the date specified in the notice is unacceptable to the owner and another date is agreed between the owner and the billing authority, that date becomes the date from which the building will be treated as complete and the original notice is deemed to be withdrawn (Sch. 4A, para. 3 (1)).

2.4.39 Similarly, a completion notice can be withdrawn by the service on the owner of another completion notice, although once the appeal procedure has been initiated, this means of withdrawal can only be carried out with the consent of the owner. (Sch. 4A para. 1 (4)). However, once the valuation tribunal has determined the completion date, this method of withdrawal is not available.

2.4.40 When a property becomes occupied, the occupied rate is payable by the occupier, at which point the owner's liability ceases.

2.4.41 A copy of the completion notice must be served on the valuation officer and, if such a notice is withdrawn or if the date specified is varied by agreement, the billing authority is required to inform the valuation officer.

2.4.42 *Structural alterations*
Where structural alterations transform a property into another hereditament or several hereditaments, the original heriditament is deemed, for the purposes of the unoccupied rate, to have ceased to exist and to have been omitted from the rating list on the date on which the structural alterations are completed (s. 46A (5) and (6)).

2.4.43 The date on which the structural alterations are deemed to be completed is to be determined in accordance with the completion notice procedure (see 2.4.28–41). However, this does not affect any liability to pay the unoccupied rate prior to the completion of the structural alterations.

2.4.44 Should the owner dispute the date given, it is open to the billing authority and the owner to negotiate another date on which the building will be deemed to be complete. But, failing agreement, the owner can appeal against the completion notice to the valuation tribunal. The appeal must, however, be made within four weeks of the service of the completion notice (reg. 29 (1) Non Domestic Rating (Alteration of Lists and Appeals) Regulations 1993 (SI 1993 No. 291) (see 9.12).

2.4.45 An appeal to the valuation tribunal must be based on the grounds that the building to which the notice relates has not been, or, as the case may be, cannot reasonably be expected to be completed by the date specified (Sch. 4A, para. 4 (1)).

2.4.46 It is also possible to ask a valuation tribunal to review its decision in circumstances where new evidence, which could not have been ascertained by reasonably diligent inquiries, or could not have been foreseen, has become available since the conclusion of the proceedings (reg. 45 (1) (b) & (6) Non Domestic Rating (Alteration of Lists and Appeals) Regulations 1993 (SI 1993 No. 291), see 9.12.2).

2.4.47 Provisions exist to ensure that owners do not avoid liability to contribute towards the cost of the post-revaluation transitional arrangements (see 4.6.4) by carrying out works which remove their property from the rating list (Non Domestic Rating (Chargeable Amounts) Regulations 1994 (SI 1994 No. 3279) – refer also to the Valuation Office Agency Practice Note – Revaluation 1995: Altered Hereditaments).

2.4.48 For further details of appeals procedures and the work of the valuation tribunal, see Chapters 9 and 10.

2.5   Check-list

2.5.1 In order for liability to the occupied rates to exist, there must be a rateable occupier. Occupation is a matter of fact and not legal title. If owners have any use of land, they are rateable as occupiers (2.2–2.3.3).

2.5.2 For there to be a rateable occupier, the four essential elements of rateable occupation must exist (2.3.4).

2.5.3 Actual possession means the physical presence of the occupier on the land or the acts by which he makes use of or controls the land. Legal title is irrelevant (2.3.5–22).

2.5.4 Exclusive occupation is the ability to exclude everyone else from using the land in the same way. Where two occupiers use land in a similar way, it is the person in paramount occupation who is the rateable occupier (2.3.23–35).

2.5.5 Beneficial occupation means that the occupation must be valuable, not profitable. The occupation must be capable of producing a rent. Land 'struck with sterility' is not rateable (2.3.36–46).

2.5.6 Permanent occupation must be considered in the light of the duration and the nature of the occupation (2.3.47–53).

2.5.7 The unoccupied rate is half the level of the occupied rate and is paid by owners of unoccupied property after a three-month 'rate-free' period (2.4).

2.5.8 Billing authorities have a discretionary power to serve a Completion Notice on owners on newly-completed or structurally-altered buildings in order to bring them into rate liability (2.4.28–46).

Chapter 3

The hereditament

3.1 Synopsis

3.1.1 The Uniform Business Rate is a tax levied on land and buildings, specifically on relevant non-domestic hereditaments, i.e., units of rateable property, being lands and buildings, mines and advertising rights.

3.1.2 Chattels may be rateable if they are occupied with and enhance the value of land. Chattels which are plant and machinery may be rateable if the items of plant and machinery are listed in the appropriate legislation. They may be valued as a hereditament together with land and buildings or as a separate hereditament.

3.1.3 The unit of rateable property which is taxed is called the 'hereditament' and is defined in case law.

3.2 Rateable property

3.2.1 Property which is liable to be rated is defined (s. 64 (4) 1988 Act) as: lands, mines, sporting rights which are severed from the occupation of the land on which the right is exercisable, and (under s. 64 (2) 1988 Act) a right to use any land for the purpose of exhibiting advertising.

3.2.2 The rating of mines and mineral rights is a specialism and is not considered further in this text.

3.2.3 Sporting rights, which were rateable prior to 1 April 1997, are now exempt (s. 2 Local Government and Rating Act, 1997 – see 4.3.17).

3.2.4   The word 'lands' is taken to mean everything on, over or under the surface of land and will, therefore, include buildings and areas of surface water (*Electric Telegraph Co. v. Salford Overseers* (1855)).

3.2.5   All land and buildings are rateable property unless specifically exempted by statute (see Chapter 4).

3.2.6   Although it is often said that a particular property is rateable, this is not strictly true. The liability to pay occupied rates rests on the occupier who is rateable in respect of his occupation of property.

3.2.7   Although not part of the land, chattels may be rated with the land, if they are enjoyed with and enhance the value of the land. For example, builders' huts and caravans are chattels, but once they are occupied with land and enhance its value, they will combine with the land to create a rateable unit (*London County Council v. Wilkins (VO)* (1957) and *Field Place Caravan Park Ltd. v. Harding (VO)* (1966)).

3.2.8   Chattels which are plant and machinery are rateable if the item of plant and machinery is specified in the relevant regulations (see 3.2.21–3.2.43). Such items are then deemed to be part of the hereditament, which is valued at its enhanced value. No account is taken of the value of any other plant and machinery which is in or on the hereditament but not specified in the regulations.

3.2.9   Certain incorporeal rights, such as easements, tolls and rights of common are rateable if the exercise and enjoyment of them involves the exclusive occupation of land. The facts of the occupation would decide liability, not the fact that the easement is the source of the right to occupy (see 2.3).

3.2.10   *Advertising rights*
Advertising rights are defined (s. 64 (2) 1988 Act) as a right to use any land for the purpose of exhibiting advertisements where:
(a)   the right is let out or reserved to any person other than the occupier of the land, or
(b)   where the land is not occupied for any other purpose, the right is let out or reserved to any person other than the owner of the land.

3.2.11   An advertising right which is not occupied together with land (and is therefore rateable under s. 64 (2) above) is considered to be occupied by the person for the time being entitled to the right (s. 65 (8) 1988 Act).

3.2.12   An advertising right may also be enjoyed together with the land and/or buildings on which it is displayed, in which case it is not rated separately from the land. Instead, the land, buildings and advertising right are valued together as one unit of rateable property, and the occupier is taxed in the normal way.

3.2.13   In cases where:
(a)   land consists of rateable property (specifically, a hereditament – see 3.3) which is used permanently or temporarily for the exhibition of advertisements or for the erection of a structure used for the exhibition of advertisements;
(b)   the right is not occupied separately from the land; and
(c)   the hereditament is not occupied apart from the advertising use;
then the hereditament is treated as occupied by the person permitting it to be so used or, if that person cannot be ascertained, by the owner (s. 65 (8A) 1988 Act).

3.2.14   *Plant and Machinery*
All plant and machinery is exempt unless it is mentioned in the Act, or the Regulations which accompany the Act.

3.2.15   The Valuation for Rating (Plant and Machinery) Regulations 1994 (SI 1994 No. 2680) were introduced with effect from 1 April 1995. The figures used for the revaluation which took effect on 1 April 1995 are those which are deemed to apply at the valuation date of 1 April 1993 (see 5.4.38 *et seq.*).

3.2.16   The 1994 regulations specify (reg. 4) that the previous 1989 regulations (Valuation for Rating (Plant and Machinery) Regulations 1989 (SI 1989 No. 441)) apply only to the 1990 rating lists, i.e., to the lists which were in force until 31 March 1995.

3.2.17   Only those items of plant and machinery specified in the Schedule to the 1994 Regulations are deemed to be part of the hereditament in or on which the item exists (reg. 2 (a) (i)). All other items of plant and machinery have no effect on the rent to be fixed (reg. 2 (a) (ii)) (including the effect of any item of plant and machinery which is not in or on the hereditament being valued (reg. 2 (b))).

3.2.18   'Plant and machinery' is not defined. In *Yarmouth v. France* (1887) (a case involving employer's negligence) 'plant' was described as

> the materials or instruments which the employer must use for the purpose of carrying on his business ... (at p. 652)

The term does not include stock-in-trade, nor does it include the place in which the business is carried on.

3.2.19   Plant and machinery may or may not be an integral part of the hereditament. It may be a separate hereditament and valued as such, or it may be a part of a hereditament and be valued together with the land and buildings, so that its value increases the value of the whole.

3.2.20   The Valuation for Rating (Plant and Machinery) Regulations 1994 (SI 1994 No. 2680) provide that in valuing any hereditament

> the value of any other plant and machinery [which is not listed] has no effect on the rent to be estimated. (reg. 2 (a))

3.2.21   Regulation 3 requires the valuation officer, on written request, to supply the occupier with particulars in writing indicating which items of plant and machinery have been assumed to form part of the hereditament or whether any particular item has been assumed to form part of the hereditament. There is no time-limit within which the valuation officer has to comply, nor is the valuation officer required to supply the valuations themselves.

3.2.22   The 1994 Regulations include a Schedule of 'Classes of Plant and Machinery to be Assumed to be Part of the Hereditament' (and, therefore, rateable).

3.2.23   Generally items are given their trade or technical names and the meaning of such terms can be proved by expert evidence. However, according to Salmon LJ, in *Chesterfield Tube Co. Ltd., v. Thomas (VO)* (1970), at p. 737:

> The question ... is what they mean to rating valuers and surveyors, the occupiers of hereditaments and the practical technicians concerned with the design, making and operation of the plant and machinery which the hereditaments contain.

3.2.24   There are four classes of plant and machinery assumed to be part of the hereditament and these are broadly grouped into power, services, conveying and named structures.

3.2.25   *Class 1 – power class*
Plant and machinery specified in Table 1 (together with any of the appliances and structures accessory to such plant or machinery and specified in the List of Accessories) which are used or intended to be used mainly or exclusively in connection with the generation, storage, primary transformation or main transmission of power in or on the hereditament.

3.2.26   'Transformer', 'primary transformation of power' and 'main transmission of power' are defined.

3.2.27   Table 1 includes steam boilers, gas turbines, storage batteries, motor generators, switchboards, sluice gates, hydraulic engines, air compressors, windmills, pulleys and wheels, motors used mainly for driving any plant and machinery falling within Class 1 and solar panels (unless used for heating water). These are all listed, and therefore rateable.

3.2.28   *Class 2 – services class*
Plant and machinery specified in Table 2 (together with any of the appliances and structures accessory to such plant or machinery and specified in the List of Accessories) which are used or intended to be used mainly or exclusively in connection with services to the land or buildings of which the hereditament consists, other than any such plant or machinery which is in or on the hereditament and is used or intended to be used in connection with services mainly or exclusively as part of manufacturing operations or trade processes.

3.2.29   'Services' is defined as heating, cooling, ventilating, lighting, draining or supplying of water and protection from trespass, criminal damage, theft, fire or other hazard.

3.2.30   Thus, the following examples from Table 2 are listed (and therefore rateable), unless they are used mainly or exclusively as part of manufacturing operations or trade processes: water heaters, radiators, refrigerating machines, electric lamps, sewage treatment plant and machinery, tanks, pumps, fire-alarm systems, blast protection walls, lightening conductors, security and alarm systems, shutters, grilles, and fences.

3.2.31   The List of Accessories relates to both Classes 1 and 2, and includes foundations, settings, gantries, supports, platforms and stagings for plant and machinery; pipes, ducts, valves, traps, separators, filters, coolers, purifying and other treatment apparatus; shafting supports, belts, ropes and chains; instruments and apparatus attached to the plant and machinery, including computers, meters, gauges, measuring and recording instruments, automatic or programmed controls, temperature indicators, alarms and relays.

3.2.32   *Class 3 – conveying class*
The following items:
(a)   railway and tramway lines and tracks and associated fixed accessories and equipment;
(b)   lifts, elevators, hoists, escalators and travelators;
(c)   cables, wires and conductors used or intended to be used in connection with the transmission, distribution or supply of electricity other than such items which are comprised in the equipment of and are situated within premises, i.e., any hereditament other than the one used or intended to be used mainly or exclusively for the transmission, distribution or supply of electricity;
(d)   poles, posts, pylons, towers, pipes, ducts, conduits, meters, switchgear and transformers, used or intended to be used in connection with any other items in (c) above;
(e)   cables, fibres, wires and conductors used or intended to be used in connection with the transmission of communication signals and which are comprised in the equipment of and are situated within premises, i.e., any hereditament which is used or intended to be used mainly or exclusively for the processing or the transmission of communications signals;
(f)   poles, posts, towers, masts, mast radiators, pipes, ducts and conduits and any associated supports and foundations used or intended to be used in connection with any of the items listed in (e) above;
(g)   a pipe-line, i.e., a pipe or system of pipes and associated fixed accessories and equipment for the conveyance of anything, not being:
(i)   a drain or sewer; or
(ii)   a pipe-line which forms part of the equipment of, and is wholly situated within, a factory or petroleum storage depot, a mine, quarry or mineral field or a natural gas storage or processing facility or gas holder site; and where a pipeline forms part of the equipment of, and is situated partly within and partly outside, such premises, excluding:
(1)   in the case of a pipe-line for the conveyance of anything to the premises, so much of the pipe-line as extends from the first control valve on the premises; and
(2)   in the case of a pipe-line for the conveyance of anything away from the premises, so much of the pipe-line as extends from the last control valve on the premises;

but not excluding so much of the pipe-line as comprises the first or, as the case may be, the last control valve.

3.2.33 *Class 4 – named structure class*
The items specified in Tables 3 and 4, except:
(a) any such item which is not, and is not in the nature of, a building or structure;
(b) any part of any such item which does not form an integral part of such item as a building or structure or as being in the nature of a building or structure;
(c) so much of any refractory or other lining forming part of any plant or machinery as is customarily renewed by reason of normal use at intervals of less than fifty weeks;
(d) any item in Table 4 the total cubic capacity of which (measured externally and excluding foundations, settings, supports and anything which is not an integral part of the item) does not exceed four hundred cubic metres and which is readily capable of being moved from one site and re-erected in its original state on another without the substantial demolition of any surrounding structure.

3.2.34 Table 3 includes the following items: blast furnaces; bridges; coking ovens; floating pontoons; foundations; fixed gantries; headgear for mines, quarries and pits; radio telescopes; settings; turntables; walkways, stairways, handrails, catwalks, stages, staithes and platforms; well casings and liners.

3.2.35 Table 4 (which is limited by the size condition imposed by (d) above) includes: accelerators; boilers; furnaces, kilns, stoves and ovens; chambers and vessels; heat exchangers; reactors; incinerators; silos; stills; tanks; vats; washeries for coal; and wind tunnels.

3.2.36 All these items must be readily capable of being moved from one site and re-erected in their original state on another site without the substantial demolition of any surrounding structure.

3.2.37 Rateable items under Class 4 must be or be in the nature of a structure or building. *Elliot's Bricks Ltd v. Hartley (VO)* (1990) provides a useful summary of all aspects that need to be considered when establishing liability for plant and machinery which is in the nature of a structure.

3.2.38 *Naming*
In order to be rateable, an item of plant and machinery must be named in the order. If it is not named in generic terms in the order, it is not rateable. Therefore, consider what the item does.

3.2.39   In *Union Cold Storage Co. Ltd. v. Phillips (VO)* (1973) it was held that cold rooms and freezer rooms, which are not listed, were, in effect, 'chambers for refrigeration' and as such listed and, therefore, rateable.

3.2.40   But it is important to remember that the trade or technical meaning will take precedence if there is any doubt if an item is listed (*Chesterfield Tube Co. Ltd. v. Thomas (VO)* (1970)). This is particularly true for items in Class 4. Such details should be requested from the occupier, the site engineer or the production engineer in the hereditament.

3.2.41   *Rateable unit*
Listed items of plant and machinery can enhance the value of a rateable unit or hereditament (see 3.3) and can be valued together with the land and buildings or form a separate rateable unit on their own.

3.2.42   Almost every hereditament contains some plant and machinery, e.g., lighting and drainage are provided by items of rateable plant and machinery, and these are normally valued as part of the property without any specific mention being made of them, regardless of the method of valuation used. Valuation officers should mention such items in order to comply with the notification requirement of reg. 3 of the 1994 regulations.

3.2.43   Items of plant and machinery which are not listed do not affect the rateable value of the hereditament and are deemed not to occupy the space they do in fact occupy. The tenant is assumed to install such non-rateable items as are required for the trade or business.

3.3   Hereditament

3.3.1   The unit of assessment for rating is the hereditament. Thus, rateable property must also be a hereditament in order to be liable to rates.

3.3.2   Under s. 64 (1) 1988 Act, a 'hereditament' is anything which, by virtue of the definition of 'hereditament' in s. 115 (1) of the General Rate Act 1967, would have been a hereditament for the purposes of that Act.

3.3.3   The pre-1990 rules must therefore be examined in order to establish the definition of a hereditament.

3.3.4   A 'hereditament' was defined in s. 115 (1) General Rate Act 1967, as:

> property which is or may become liable to a rate, being a unit of such property which is, or would fall to be, shown as a separate item in the [rating] list.

3.3.5   This means that each hereditament has a separate entry in the list in which all hereditaments are itemised. To define a 'hereditament' as being a unit of rateable property which is a separate entry in the rating list presupposes a knowledge of what a hereditament is before it can be entered into the rating list.

3.3.6   In other words, if rateable property is to be a hereditament, it must be a separate entry in the rating list; but it cannot be a separate entry until it is a hereditament! There is therefore no statutory guidance as to its meaning.

3.3.7   Fortunately, it is possible to refer to case law which defines a hereditament as an item of rateable property which is:
(a)   capable of definition;
(b)   a single geographical unit;
(c)   capable of separate occupation;
(d)   put to a single use.

3.3.8   Relevant non-domestic property which does not conform to the above is not a hereditament and is not, therefore, liable for rates. Each of these points will now be considered in detail.

3.3.9   **Capable of definition**
It must always be possible to identify the actual land (or rateable property) which is to be assessed, even if no physical boundaries exist.

3.3.10   Thus, where a golf club had a licence to use an undefined area of the New Forest as a golf club, it was held that the club was not in rateable occupation of any hereditament (*Harding (VO) v. Bramshaw Golf Club* (1960)).

3.3.11   Similarly in *Spear v. Bodmin Union* (1880), where two stalls were rented in a market but no particular site laid out for them, the stall occupier was not rateable, because it was impossible to identify the land he occupied.

3.3.12   However, if the land on which the stall had been placed had been capable of definition, then, subject to a sufficient degree of permanence (see 2.3.47 *et seq*), the land together with the stall (which was a chattel) would have combined to form rateable property and, therefore, a hereditament capable of assessment.

3.3.13   *Single geographical unit*
A hereditament must form a single geographical unit, so that two properties which are divided by property in the occupation of another will always be two separate hereditaments (but see *Gilbert (VO) v. Hickinbottom & Sons Ltd.* (1956) at 3.3.29–3.3.33).

3.3.14   Where two properties in the same occupation are contiguous (and conform to the other criteria), they may be one hereditament.

3.3.15   'Contiguous' in this context, means 'touching', not 'adjacent' or 'close'. It should be possible to draw a notional 'ring-fence' or 'three-dimensional envelope' around a hereditament without including property which is (rateably) occupied by another.

3.3.16   Premises on the opposite sides of a road will be contiguous if the highway is in the exclusive occupation of the occupier of the premises (but see *Gilbert (VO) v. Hickinbottom & Sons Ltd.* (1956) at 3.3.29–3.3.33). Connections by cables, pipes, wires or railway lines will not create contiguity.

3.3.17   In one case, a university occupied a number of buildings beyond the main campus. Although functionally part of the university, the buildings were scattered throughout the town, and because they were dispersed among the buildings in the occupation of others, they were each held to be separate hereditaments. Each separate property formed a separate hereditament, all in the occupation of the university. (*University of Glasgow v. Assessor for Glasgow* (1952) – although a Scottish case and not binding on English law, it is considered to be persuasive.)

3.3.18   A functional connection between two non-contiguous properties may cause them to be treated as a single hereditament (see 3.3.29 *et seq.*).

3.3.19   *Separate occupation*
Where one property is capable of being separately occupied, for example, a single shop occupied by a single business unit – that is a single hereditament.

3.3.20   Where property is capable of being separately occupied by several occupiers, for example, a block of six self-contained shop units, each shop is a separate hereditament and thus the block comprises six separate hereditaments.

3.3.21 Where one person occupies contiguous properties, the several parts of which are capable of being occupied separately by several occupiers, they will be treated as one hereditament, unless there is some special reason why they must be treated as several hereditaments (e.g., if they are put to different uses (see 3.3.23 *et seq.*)).

3.3.22 Thus, several units in a parade of shops, all occupied by the same person, may be a single hereditament, if they are contiguous although not internally connecting, providing all the other criteria are met.

3.3.23 *Single purpose*
Premises in one occupation may form two or more hereditaments if parts of the premises are used for wholly different purposes.

3.3.24 Premises which are contiguous and which may form a single hereditament if in the same occupation will become separate hereditaments if they are used for substantially different purposes.

3.3.25 For example, in *North Eastern Rail. Co. v. York Union* (1900), it was held that the hotel and refreshment rooms should be separately rated from engine sheds, carriage and wagonshops, etc., pumping station, coal yards and warehouses, all in the occupation of the railway company.

3.3.26 In *Burton Latimer Urban District Council v. Weetabix Ltd. & Lee (VO)* (1958), a warehouse and sports ground, both used in connection with an adjoining factory, were held to form one hereditament.

3.3.27 However, in the days when domestic property was rated, it was established that where living accommodation adjoined and was occupied with business premises and was used in connection with the business, it was rated as a single hereditament with the non-domestic premises. This is now no longer the case (see Composite hereditaments, 3.3.37 *et seq.*).

3.3.28 Premises which would otherwise be separate hereditaments may form a single hereditament if there is a sufficiently strong functional connection between them.

3.3.29 In *Gilbert (VO) v. Hickinbottom & Sons Ltd.* (1956), the Court of Appeal considered whether two premises in one occupation but divided by a public highway constituted a single hereditament. One of the premises was a retail bakery and the other a repair depot, situated immediately opposite each other, either side of a highway. The repair depot was used primarily to repair and maintain the bakery's commercial vehicles, but was also used to repair any breakdown in the bakery's plant and machinery.

3.3.30 It was found to be essential that any breakdown in the bakery plant should be dealt with immediately to avoid any interference with production and for this reason the repair depot was constantly manned by an engineer or electrician. The Court of Appeal held that the two premises formed a single hereditament.

3.3.31 If several buildings in one occupation are physically separated, being interspersed among buildings in other occupations, they cannot be a single hereditament (see 3.3.13–3.3.18). It is only because the road was not in the rateable occupation of anyone that the functional test could be applied.

3.3.32 These principles have been applied by the Lands Tribunal in many cases where premises in one occupation were divided by a public highway. Those where the functional connection was held to be strong enough to overcome the geographical separation of the premises appear to fall into two categories:
(a) where the process of manufacture proceeds in both parts; and
(b) where manufacture by a continuous process goes on in one part and what goes on in the other is essential to prevent breakdown.

3.3.33 The Lands Tribunal has referred to the analogy of a sparking plug:

> where the gap between the two parts is so small that it can physically be traversed in the course of the functioning of the whole. It might also be true to say that the stronger the spark, the greater the gap which can be traversed. (*Edwards (VO) v. B. P. Refinery (Llandarcy) Ltd.* (1974) at p. 48)

3.3.34 **Hereditaments divided by local authority boundaries**
Under the pre-1990 legislation, a hereditament needed to exist within the area of a rating authority (billing authority). This was because each local authority was entitled to spend the revenue raised from property located within its boundaries.

3.3.35 This is no longer the case because the Uniform Business Rate is a central government tax which is redistributed to billing authorities on a *per capita* basis of the number of residential occupiers. Therefore, if a property straddles a boundary between two local authorities, there is no need to apportion the property value or the tax raised between the authorities. Administrative convenience takes precedence over an authority's concern for an apparent equitable split of value.

3.3.36 Thus, where a unit of property would, but for being divided by a boundary between billing authorities, fall to be treated as a single hereditament, it must be treated as situated in the area in which it appears to the valuation officer(s) for the authorities concerned to have the greater rateable value. Where they cannot agree, they must decide by lot (Non Domestic Rating (Miscellaneous Provisions) Regulations 1989 (SI 1989 No. 1060) reg. 6(5)).

3.3.37 *Composite hereditaments*
Only a relevant non-domestic hereditament is liable for rates, as long as at least part of it is neither domestic nor exempt (see Chapter 4). Liability to the Uniform Business Rate therefore depends on identifying a 'relevant non domestic hereditament' and entering that into the appropriate rating list (see Chapter 8).

3.3.38 A hereditament is non-domestic if it is either:
(a) composed entirely of property which is non-domestic; or
(b) a composite hereditament.

3.3.39 'Domestic property' is defined (s. 66 1988 Act) as property used wholly for the purposes of living accommodation (see 4.3.1 for a full definition of 'domestic').

3.3.40 A 'composite hereditament' is defined (s. 64 (9) 1988 Act) as a hereditament part of which consists of domestic property.

3.3.41 Thus, a ground-floor shop occupied together with first-floor living accommodation is a composite hereditament, as is a multi-storey office block incorporating a caretaker's flat.

3.3.42 The non-domestic element of the property is liable to rates and the whole property appears in the rating list, identified as a composite hereditament. There is a liability to the Council Tax in respect of the occupation of the domestic element of the property (see Chapter 13). See also 5.3.26–5.3.33 for the method of valuation for composite hereditaments.

3.4 Occupation of premises where only part is used

3.4.1 Generally, if premises are only partly used, then rateability can only be avoided by giving up occupation of the whole property.

(Occupation of part is occupation of whole.) However, the Lands Tribunal decision in *Moffatt (VO) v. Venus Packaging Ltd.* (1977) has accepted that structural severance is not essential to the existence of two hereditaments, even when one 'hereditament' is vacant and both are in the same ownership.

3.4.2   Under s. 44A (1988 Act), in certain cases the billing authority can initiate steps for a reduction in rateable value of a hereditament which is partly unoccupied, if it appears to the authority that part of the hereditament is unoccupied but will remain so for a short time only.

3.4.3   Under such circumstances, the authority may require the valuation officer to apportion the rateable value of the hereditament between the occupied and unoccupied parts and rates will be paid on the value apportioned to the occupied part.

3.4.4   The reduction comes to an end when any unoccupied part is re-occupied, or when a further apportionment of value under these provisions takes effect.

3.4.5   There is no provision for a ratepayer to claim or enforce a reduction if the billing authority rejects the ratepayer's request or disagrees with the valuation officer's apportionment.

3.5   Regulations

3.5.1   The Secretary of State for the Environment may make regulations under s. 64 (3) 1988 Act, providing that, in prescribed cases:
(a)   anything which would (apart from the regulations) be one hereditament shall be treated as more than one hereditament; and
(b)   anything which would (apart from the regulations) be more than one hereditament shall be treated as one hereditament.

3.5.2   Under the Non Domestic Rating (Caravan Sites) Regulations 1990 (SI 1990 No. 673), pitches on caravan sites which are not occupied by the site occupier are treated, together with any area of the site occupied by that occupier, as one hereditament and as occupied by him (reg. 3 (1)).

3.6   Check-list

3.6.1   Property which is liable to rates comprises land and buildings, mines, and advertising rights (3.2.1–3.2.13).

3.6.2   Chattels may be rateable if they are occupied with land and enhance its value (3.2.7).

3.6.3   Plant and machinery are rateable if the items are listed in statute. They can be rateable together with land and buildings or form a separate hereditament (3.2.14–3.2.43).

3.6.4   In order to be liable to rates, rateable property must also be a hereditament, defined by statute as being a separate item in the rating list (3.3.4–3.3.6).

3.6.5   Case law requires a hereditament to be:
(a)   capable of definition (3.3.9–3.3.12);
(b)   a single geographical unit (3.3.13–3.3.18);
(c)   capable of separate occupation (3.3.19–3.3.22); and
(d)   used for a single purpose (3.3.23–3.3.33).

3.6.6   Hereditaments divided by the boundary of a local authority are treated as existing entirely within one of local authority areas (3.3.34–3.3.36).

3.6.7   A hereditament is a 'composite hereditament' if it consists, in part, of domestic property (3.3.37–3.3.42).

3.6.8   Where part only of a hereditament is occupied, the billing authority has discretion to request an apportionment of the rateable value and to levy rates only on the value ascribed to the occupied part (3.4).

3.6.9   The Secretary of State has the power to prescribe that property which would be two or more hereditaments should be treated as one hereditament and vice versa (3.5).

Chapter 4

Exemptions and reliefs

4.1 Synopsis

4.1.1 All land and buildings are rateable, unless statute says otherwise.

Exemption from rates can be achieved by:
(a) exempting the property from being liable to assessment, so that the occupier (or owner) has no liability to pay rates;
(b) exempting the occupier from liability to rates in respect of the hereditament occupied;
(c) altering the rules affecting the valuation of a particular class of hereditament; or
(d) giving the billing authority the power to reduce or remit rates normally recoverable.

4.2 Legislation

4.2.1 Exemptions and relief from rates are contained in ss. 47–9 and 51 and Sch. 5 of the 1988 Act, with transitional arrangements, phasing-in increases and decreases in rate liability, being introduced following the 1990 and 1995 revaluations.

4.2.2 'Exempt' means exempt from local non-domestic rating. Any land or building not in use or occupied is treated as used in a particular way if it appears that when next in use or occupied it will be used or occupied in that way and a person is treated as an occupier of any land or building which is not occupied if it appears that when it is next occupied he will be an occupier of it (Sch. 5 para. 21).

4.2.3 The Secretary of State has the power to make regulations prescribing hereditaments to which exemption or a lesser degree of relief from rates may apply (Sch. 5, para. 20).

4.3   Exempted property

4.3.1   *Domestic property*
A hereditament which is entirely domestic property is not entered into the rating list, and is not liable to rates, but is liable to the Council Tax, with effect from 1 April 1993 (see Chapters 13–15 for Council Tax). A hereditament which is partly domestic property is a composite hereditament (see 3.3.37). A composite hereditament is rateable and entered and identified in the rating list as such (s. 42 (2) 1988 Act) (see Chapter 8).

4.3.2   Property is domestic (according to s. 66 (1) 1988 Act, as amended) if it is:

(a)   used wholly for the purposes of living accommodation;

(b)   a yard, garden, outhouse or other appurtenance belonging to or enjoyed with property used wholly for the purposes of living accommodation;

(c)   a private garage used wholly or mainly for the accommodation of a private motor vehicle;

(d)   private storage premises used wholly or mainly for the storage of articles of domestic use;

(e)   a mooring, if it is occupied by a boat which is the sole or main residence of an individual (s. 66 (4) 1988 Act);

(f)   a caravan pitch, if it is occupied by a caravan which is the sole or main residence of an individual (s. 66 (3) 1988 Act);

(g)   short-stay accommodation which is provided for short periods to less than six individuals simultaneously whose sole or main residence is elsewhere and which is self-contained, self-catering accommodation provided commercially and where the 'owner' is resident (s. 66 (2) (2A) 1988 Act);

(h)   self-catering accommodation available for letting commercially for short periods totalling less than 140 days (s. 66 (2B) (2C) 1988 Act).

4.3.3   'Private garage' is defined (s. 66 (1) (c) 1988 Act) as either a building having a floor area not exceeding 25 square metres, and which is used wholly or mainly for the accommodation of a motor vehicle (and for this purpose 'building' includes part of a building).

4.3.4   'Private storage premises' are defined in s. 3 (4)(c) Local Government Finance Act 1992 as part of a dwelling used wholly or mainly for the storage of articles of domestic use.

4.3.5   Car-parking spaces within the curtilage of, or let with, domestic property are treated as domestic and therefore exempt. Similarly, a space within the curtilage of, or let with, a non-domestic property is non-domestic, as is a free-standing car-parking space. Car parks are domestic when they are within the curtilage of blocks of flats or allocated solely to the use of the occupiers of those flats.

4.3.6   Property not in use is domestic if it appears that, when next in use, it will be domestic (s. 66 (5) 1988 Act). (Note the use of the word 'property', as opposed to 'hereditament'. It follows that part only of a 'hereditament' may be 'domestic property', e.g., a caretaker's flat in an office building.)

4.3.7   The Secretary of State for the Environment may order, amend or substitute another definition for any definition of 'domestic property' for the time being effective for the purposes of non-domestic rating (s. 3 (6) 1992 Act).

4.3.8   *Agricultural land and buildings*
Agricultural land and agricultural buildings are exempt (Sch 5, para. 1, 1988 Act).

4.3.9   Agricultural land is defined (Sch. 5, para. 2, 1988 Act) as land used as arable, meadow or pasture ground; land used for a plantation or a wood or for the growth of saleable underwood; land exceeding 0.10 hectare and used for the purposes of poultry farming; anything which consists of a market garden, nursery ground, orchard or allotment; or land occupied with, and used solely in connection with the use of a building which (or buildings each of which) is an agricultural building.

4.3.10   Agricultural land does not include land occupied together with a house as a park; gardens (other than market gardens); pleasure grounds; land used mainly or exclusively for purposes of sport or recreation; or land used as a racecourse (Sch 5, para. 2, 1988 Act).

4.3.11   Agricultural buildings are also exempt (Sch. 5, para. 1, 1988 Act), and 'building' includes a separate part of a building (Sch. 5, para. 8(4) 1988 Act).

4.3.12   A building is an agricultural building if it is occupied together with agricultural land and is used solely in connection with agricultural operations on the land; or if it is or forms part of a market garden and is used solely in connection with agricultural operations at the market garden (Sch 5, para. 3, 1988 Act).

4.3.13   'Occupied together with' means a functional rather than geographical connection, and it is not therefore necessary to have an agricultural building situated on the agricultural land with which it is occupied. However, it must be possible to identify agricultural land together with which the buildings are solely used.

4.3.14   In determining whether a building used in any way is solely so used, no account shall be taken of any time during which it is used in any other way, if that time does not amount to a substantial part of the time during which the building is used (Sch 5, para. 8 (3) 1988 Act).

4.3.15   Buildings used for the keeping and breeding of livestock, and ancillary buildings (Sch. 5, para. 5 (1) 1988 Act), and certain buildings occupied by a corporate body and used in connection with agricultural operations on agricultural land are also agricultural buildings (Sch. 5 para. 4 1988 Act).

4.3.16   Despite this rather sweeping exemption, certain uses often associated with agriculture are rateable: e.g., estates offices, estates yards, stud buildings, gallops, mineral rights, converted farm buildings and land used mainly for sport. In addition, agricultural dwelling-houses are liable to Council Tax (see Chapters 13–15).

4.3.17   **Sporting rights**
Sporting rights are defined (s. 64 (4) (d) 1988 Act) as any right of fowling, of shooting, of taking or killing game or rabbits, or of fishing when severed from the occupation of the land on which the right is exercisable. They are exempt with effect from 1 April 1997 (s. 2 Local Government and Rating Act 1997).

4.3.18   Where sporting rights are not severed from the occupation of the land on which they are exercisable, the increase in value to the land attributable to the sporting right is ignored and the land is valued without the benefit of the sporting rights.

4.3.19   Where a sporting right is severed from the occupation of the land on which it is exercisable and the severance is effected by deed, then it becomes a separate (incorporeal) hereditament, exempt rates with effect from 1 April 1997, under s. 2 Local Government and Rating Act 1997.

4.3.20 Prior to 1 April 1997, where sporting rights were not severed from the occupation of the land on which they were exercisable, the land was valued with the benefit of sporting rights, In such cases, therefore, the sporting right was valued together with the land. If the land was exempt (e.g., agricultural land – see 4.3.8–4.3.16), the sporting right was also exempt.

4.3.21 Prior to 1 April 1997, it was the owner of the sporting right who was rateable and, for this purpose, 'owner' was defined (s. 65 (9) 1988 Act) as the person entitled to receive rent (if the right was let) or the person entitled to exercise the right to let (if the right was not let).

4.3.22 Considerations which would affect the value of sporting right have been held (*Myddleton v. Charles (VO)* (1957)) to be:
(a)   the general nature of the country;
(b)   the condition of coverts and woodlands;
(c)   the extent of disturbing influences, such as roads, public footpaths; and
(d)   the facilities for proper keepering and the degree of vermin infestation which, while not permanent, is likely to affect the head of game for a season or two.

4.3.23 There are about 6,000 sporting rights in England which, prior to 1 April 1997, were separate hereditaments. They were estimated to have an aggregate rateable value of £8.2 million and to be responsible for less than £5 million in rates paid.

4.3.24 *Fish farms*
Land which is used solely for or in connection with fish-farming, and buildings (other than dwellings) so used are exempt rates (Sch. 5, para. 9 (1) 1988 Act). 'Fish-farming' means the breeding or rearing of fish, or the cultivation of shellfish, for the purpose (or for purposes which include) transferring them to other waters or producing food for human consumption (ibid. (4)). 'Shellfish' includes crustaceans and molluscs of any description but excludes ornamental fish (ibid. 4A)(5)).

4.3.25 *Fishing rights*
Fishing rights which constitute a hereditament are exempt if they are exercisable in certain fisheries (Sch. 5, para. 10, 1988 Act) (see also Sporting rights, 4.3.17).

4.3.26   *Places of public religious worship*
Places of public religious worship together with a church hall, chapel hall or similar building used in connection with a place of public religious worship are exempt rates (Sch. 5, para. 11).

Also exempt are hereditaments occupied by an organisation responsible for the conduct of public religious worship and used for carrying out administrative or other activities relating to the organisation of the conduct of public religious worship in such a place (ibid.).

4.3.27   *Property in enterprise zones*
A hereditament is exempt (Sch. 5, para. 19, 1988 Act) to the extent that it is situated in an enterprise zone. During that time, hereditaments are not entered into the rating list nor are they valued. The exemption ceases once the designation of the enterprise zone expires.

4.3.28   Once the designation of an enterprise zone ceases, the rateable properties are valued and entered into the appropriate rating list(s). Under the Valuation for Rating (former Enterprise Zones) Regulations 1995 (SI 1995 No. 213), the existence of enterprise zones is to be disregarded in making or altering a rateable value in relation to a hereditament (or part of a hereditament) which was situated in the former zone.

4.3.29   *Miscellaneous exempt property*
Also exempt are:
(a)   air-raid-protection works which are not occupied or used for any other purpose (Sch. 5 para. 17, 1988 Act);
(b)   a mooring which is used or intended to be used by a boat or ship and which is equipped only with a buoy attached to an anchor, weight or other device and which rests on the sea- or river-bed and which is designed to be raised from time to time (Sch. 5 para. 18, 1988 Act);
(c)   sewers and accessories to sewers (Sch. 5 para. 13, 1988 Act);
(d)   properties of drainage authorities (Sch. 5 para. 14, 1988 Act);
(e)   road crossings over watercourses (Sch. 5 para. 18A, 1988 Act);
(f)   a park which has been provided by, or is under the management of, a local authority and is available for free and unrestricted use by members of the public (Sch. 5 para. 15, 1988 Act) (see also 2.3.42);
(g)   empty industrial hereditaments, including so-called 'moth-balled' factories and warehouses (see 2.4.16–2.4.20).

4.4 Exempted occupiers

4.4.1 *Diplomatic occupations*
Diplomats and persons with diplomatic immunity are not subject to the processes of British law and, therefore, should they perform an illegal act, they cannot be prosecuted in the British courts.

4.4.2 Thus, if a diplomat, etc., fails to pay a rates bill which is served on him, then there is no remedy through the courts to enforce that payment. In practical terms, therefore, such individuals are exempt rates. However, diplomatic missions normally pay a 'beneficial proportion' of the rates otherwise payable.

4.4.3 With effect from April 1991, the Valuation Office Agency is responsible for the collection of a beneficial portion of the rates on diplomatic property, as well as for recovery of such a portion of rates from foreign missions and international organisations.

4.4.4 *Properties used for the disabled*
A hereditament is exempt provided it is used wholly for the provision of facilities and for the provision of welfare services and training, etc., for disabled persons (Sch. 5 para. 16, 1988 Act).

4.4.5 *Trinity house property*
The following kinds of property, owned or occupied by Trinity House are exempt: buoys; beacons; lighthouses and property within the same curtilage as, and occupied for the purpose of, a lighthouse (Sch. 5, para. 12, 1988 Act).

4.5 Discretionary power to remit rates

4.5.1 *Charities and other non-profit-making organisations*
A combination of mandatory and discretionary relief is available for charitable and non-profit-making occupations (s. 47 1988 Act).

4.5.2 Mandatory relief of 80% is available to property wholly or mainly used for charitable purposes which is occupied by a registered charity. At the billing authority's discretion, additional relief of up to 20% is available to such occupiers.

4.5.3 For occupiers which are not registered charities, discretionary relief of up to 100% is available at the discretion of the billing authority to property all or part of which is occupied for the purposes of a non-profit-making organisation whose main objectives are philanthropic or religious or concerned with education, social welfare, science, literature or the fine arts; or a non-profit-making club, society or other organisation, which is used for the purposes of recreation (s. 47(2) 1988 Act).

4.5.4   Relief from rates is extended to 'charity shops', specifically, premises used wholly or mainly for the sale of goods given to a charity, provided the proceeds of sale are applied to the purposes of the charity.

4.5.5   Three-quarters of the cost of any relief is borne by the national non-domestic rating pool.

4.5.6   A practice note (*Non-Domestic Rates: Discretionary Rate Relief*, Department of the Environment and the Welsh Office, 1989) sets out some of the criteria which billing authorities should take into account in the exercise of their discretion. The criteria are not intended as a rigid set of rules, but describe good practice in the consideration of individual cases.

4.5.7   *Village shops*
The Local Government and Rating Act 1997 (S.1 and Sch. 1) has introduced provisions to permit billing authorities to grant up to 100% relief to any shop or public house provided that:
(a)   it is within a rural settlement of 3,000 people or less; and
(b)   it has a rateable value of less than the amount prescribed by the Secretary of State by order; and
(c)   the billing authority is satisfied that the hereditament is of benefit to the community; and
(d)   the billing authority considers it reasonable to grant relief, having regard to the interests of its Council-Tax payers.

To date, no regulations have been made which specify the relevant rateable value.

4.5.8   *Hardship*
Section 49 of the 1988 Act gives billing authorities the power to reduce or remit the payment of rates on the grounds of 'hardship'.

4.5.9   However, an authority may not give relief unless it is satisfied that the ratepayer would otherwise sustain hardship and it is reasonable to do so having regard to the interests of its council-tax payers.

4.6   Altering the rules for fixing rateable value

4.6.1   It is possible for central government to vary the normal rules which are applied to the assessing of rateable value, and it has done so for those hereditaments occupied by gas, water, docks and harbours, electricity supply and railway undertakers (see 6.3).

4.6.2   However, there are also rules which alter a normally assessed rateable value to provide a value on which rates are payable and this currently applies to stud farms and to those hereditaments affected by transitional arrangements.

4.6.3   *Stud farms*
Stud farms receive a degree of relief from the payment of rates, under Sch. 6 para. 2A, 1988 Act and the Non-Domestic Rating (Stud Farms) Order 1989 (SI 1989 No. 2331) The level of relief takes the form of a deduction from the rateable value of the lesser of:
(a)   £2,500; and
(b)   the rateable value.

4.6.4   *Transitional arrangements*
Prior to the 1995 revaluation taking effect, central government attempted to minimise the combined effects of the Uniform Business Rate and the first revaluation for 17 years (which took effect in 1990) by phasing in both increases and decreases in rate liability.

4.6.5   The Treasury decided that any transitional arrangements should be self-financing. Thus, the phased increases in rates for those who would otherwise suffer major increases in rate liability ('losers') were paid for by those occupiers whose rate liability decreased ('gainers') and who had their reduced rate liability phased in to pay for it.

4.6.6   Transitional arrangements for the 1995 list, which continue this philosophy (and which are fiendishly complicated), are contained in the Non Domestic Rating (Chargeable Amounts) Regulations 1994 (SI 1994 No. 3279). They apply to hereditaments which have a rateable value on 31 March 1995 and a rateable value on 1 April 1995, regardless of whether they are occupied on those dates. Transitional arrangements are not, therefore, available for hereditaments which are entered into the rating list for the first time on or after 1 April 1995 (see 2.4.47).

4.6.7   The arrangements continue to apply following a change of occupier and for the relevant period, i.e., the financial years covered by the 1995 rating list (1995–96, 1996–97, 1997–98, 1998–99 and 1999–2000) (reg. 2, ibid.).

4.6.8   Transitional arrangements apply to a 'defined hereditament', i.e., any hereditament existing in the lists during the relevant period (see 4.6.7), and which was also entered in the previous 1990 rating list.

4.6.9   Initially, it is necessary to calculate the amount of rates which would have been paid had there been no transitional arrangements, and this is called the 'notional chargeable amount' (reg. 4. ibid.).

4.6.10   Next, it is necessary to calculate the 'base liability', which varies according to the transitional arrangements which applied to the hereditament under the 1990 rating list (regs. 5–7 ibid.).

4.6.11   Then, it is necessary to calculate the 'appropriate fraction' (reg. 8 ibid.).

4.6.12   Basically, the chargeable amount (i.e., the amount of rates actually payable) is determined by multiplying the base liability by the appropriate fraction.

4.6.13   The regulations cover a multitude of possible situations, e.g., hereditaments in the central list; where a hereditament is merged or split, and all inquiries as to the details of the transitional arrangements should be made to the regulations.

4.6.14   Table 4.1 is, however, a much simplified version of how the transitional arrangements affect properties.

4.6.15   Remember that the percentages which applied to the so-called 'gainers', i.e., the decreases, were varied annually as the cost of paying for the support given to the so-called 'losers', i.e., the phased-in increases, is known.

4.6.16   The provisions of the 1996 budget have frozen the transitional arrangements for the so-called 'losers' who were having their full 1995 rate liability phased in and permitted the so-called 'gainers' to benefit immediately from the lower level of rates which were payable following the 1995 revaluation.

4.7   Reform

4.7.1   There continues to be debate about the nature and the extent of exemptions and reliefs from the payment of UBR. The debate surrounding the exemptions given to occupiers of agricultural land and buildings has continued for decades.

4.7.2   The reform introduced by S. 3 of the Local Government and Rating Act 1977, has removed the exemption from Crown occupation, so that all occupations by the Crown, including occupation by servants of the Crown (e.g., army barracks and government offices) and by those who perform duties on behalf of the Crown, e.g., the general administration of the country and which was also a public purpose of central government are now rateable.

4.7.3   Indeed, a recommendation (recommendation 56) of the Bayliss Report (see Appendix C) is that the whole issue of exemptions should be the subject of a government committee of inquiry.

4.7.4   It can be argued that there is little justification for granting relief from the payment of a tax based on the value of land for any occupier and that any relief should be funded by central government. Similarly, the justification for phasing in increases and decreases in rate liability following a revaluation is extremely tenuous, when the whole reason for the revaluation is to redistribute the rate-liability based upon up-to-date property values.

4.7.5   Nevertheless, this is probably a debate best left until the rating system can be considered as a whole. The criticisms of the rating system are considered in Chapter 12.

4.8   Check-list

4.8.1   Domestic property (as defined) is exempt rates but its occupiers are liable for Council Tax (see 4.3.1–4.3.7 and Chapters 13–15).

4.8.2   Agricultural land and buildings (as defined) are exempt rates (4.3.8–4.3.16).

4.8.3   Sporting rights are exempt (4.3.17–4.3.23).

4.8.4   Fish farms are exempt rates (4.3.24).

Table 4.1 Transitional arrangements

Maximum annual increases in real terms		
Year	**Small properties**	**Other hereditaments**
1995–96	7.5%	10%
1996–97	7.5%	10%
1997–98	7.5%	10%
1998–99	7.5%	10%
1999–2000	7.5%	10%

Maximum annual decreases in real terms		
Year	**Small properties**	**Other hereditaments**
1995–96	10%	5%
1996–97	10%	5%
1997–98	20%	15%
1998–99	35%	30%
1999–2000	35%	30%

'Small Properties' are those with a rateable value of less than £15,000 in Greater London and with a rateable value of less than £10,000 elsewhere.

Note that the annual decreases for 1997–98 onwards are estimates only. They will be reviewed annually in the light of the cost of supporting the phased-in increases.

4.8.5   Certain fishing rights are exempt (4.3.25).

4.8.6   Places of public religious worship, together with church halls, etc., are exempt (4.3.26).

4.8.7   Property in an enterprise zone is exempt (4.3.27–4.3.28).

4.8.8   Air-raid-protection works, 'swing' moorings, sewers, the property of drainage authorities, local authority parks, and empty industrial hereditaments are also not rateable (4.3.29).

4.8.9   Diplomatic occupations are exempt rates, but a proportion of the amount otherwise due is paid (4.4.1–4.4.3).

4.8.10   Properties used by the disabled are exempt (4.4.4).

4.8.11   Certain property occupied by Trinity House is exempt (4.4.5).

4.8.12   Mandatory relief of 80% is available to registered charities, which, at the discretion of the billing authority, can be increased to 100% of the rates otherwise payable (4.5.2).

4.8.13   For other non-profit-making organisation, the billing authority has discretion to grant relief of up to 100% of the rates otherwise payable (4.5.3–4.5.6).

4.8.14   Billing authorities have power to grant up to 100% relief to so-called village shops and public houses (4.5.7).

4.8.15   Billing authorities have discretionary powers to reduce or remit rates on the grounds of hardship (4.5.8–4.5.9).

4.8.16   The rateable values of stud farms are reduced by a maximum of £2,500 (4.6.3).

4.8.17   Transitional arrangements are available to reduce the impact of rates following a revaluation (4.6.4–4.6.16)

4.8.18   See Chapter 12 and Appendix C (The Bayliss Report) for a consideration of the appropriateness of the current exemptions and reliefs applied to the UBR (4.7).

Chapter 5

Rateable value

5.1   Synopsis

5.1.1   The basis of assessment is rateable value (RV) which, when multiplied by the UBR for any year, gives the amount of rates payable. Thus, rateable value (RV) $\times$ UBR = rates payable.

5.1.2   Rateable value is a net annual rent payable by a hypothetical tenant for the property, assuming certain tenancy conditions, as at the valuation date.

5.1.3   Rateable value is fixed assuming certain principles which relate to all property types, including an antecedent valuation date of 1 April 1993.

5.2   Legislation

5.2.1   For each day on which a hereditament is shown in the local rating list, there must also be shown a rateable value (RV), either for the whole property or for the part of the property which is neither exempt nor domestic property (s. 42 (4) 1988 Act).

5.2.2   The rates actually demanded by the billing authority are levied by reference to the rateable value of a hereditament, thus: rates paid = rateable value (RV) $\times$ UBR.

5.2.3   The level of the UBR is fixed by the Secretary of State for the Environment (and the Secretary of State for Wales) in 1990–91 and in 1995–96 (and for each financial year in which a revaluation takes effect), and thereafter by the Chancellor of the Exchequer (see 1.6).

5.2.4   The rateable value is fixed by the valuation officer who is responsible for the rating list (see Chapter 8) for the billing authority area in which the hereditament is situated.

5.2.5   Valuation officers are, since September 1991, employed by the privatised Valuation Office Agency, which was previously part of the Department of Inland Revenue.

5.2.6   All rateable values for hereditaments within the area of a billing authority are contained in the local non-domestic rating list held by the valuation officer, a copy of which is held by the billing authority (see Chapter 8).

5.3   Rateable value

5.3.1   Rateable value is defined (Sch. 6, para. 2 (1) 1988 Act) as:

... an amount equal to the rent at which it is estimated the hereditament might reasonably be expected to let from year to year if the tenant undertook to pay all usual tenant's rates and taxes and to bear the cost of the repairs and insurance and the other expenses (if any) necessary to maintain the hereditament in a state to command that rent.

5.3.2   Rateable value is, therefore, a net annual rent, assuming that the tenant is responsible for all outgoings.

5.3.3   *Hypothetical tenancy*
The statutory definition of rateable value requires the assessment of a rent, assuming that the tenant who pays that rent has certain specified responsibilities. This mention of a tenant gives rise to the assumption that a tenancy has been created.

5.3.4   Because the tenancy does not in fact exist, it is called the 'hypothetical tenancy' and the tenant and landlord involved are the 'hypothetical tenant' and the 'hypothetical landlord'.

5.3.5   Thus, the definition of rateable value gives rise to the use of the terms 'hypothetical tenancy' and the 'hypothetical tenant' and the characteristics of each are relevant to the assessment of rateable value. Although neither of these terms is found in statute, they can be defined by a study of case law.

5.3.6   *Hypothetical tenant*
The statutory definition requires the assumption that the property

is to be rented to a tenant. Therefore, the fact that it is in reality occupied is disregarded.

5.3.7 All possible occupiers, including the actual occupier, must be considered as possible tenants, on a year to year basis. As a result of the House of Lords decision in *R. v. School Board for London* (1886), even though it may not be possible in fact and may be forbidden by law that the actual occupier should be a tenant of a hereditament, it must be supposed that the actual occupier is among the potential tenants, for the purpose of valuing that hereditament for rating. Similarly, an owner not in occupation can be regarded as a potential hypothetical tenant.

5.3.8 The House of Lords took the principles in the above case further in *Davies v. Seisdon Union* (1908). Where a sewage board let a sewage farm to a tenant at a rent higher than that which the farm would have commanded without the manurial value of the sewage, it was held that the board should be included among the possible hypothetical tenants as this rent alone was not necessarily the measure of rateable value, and that it would have been willing to pay a rent for the farm in order to carry out its statutory duty.

5.3.9 The rent to be estimated is the rent which 'might reasonably be expected' (i.e., not the actual rent paid nor the legally demandable rent) for the hereditament if 'let from year to year' and not for a term of years. However, it must be assumed that there is a reasonable prospect of continuance, and the 'year to year' rent must be calculated on that basis (*R. v. South Staffordshire Waterworks Co.* (1885)).

5.3.10 Although the hypothetical tenancy must be deemed to have a reasonable prospect of continuance, this does not mean that all events anticipated to take place in the foreseeable future will affect the rental value of the hereditament, although most will necessarily affect the annual rent.

5.3.11 Only those events which affect the annual rental bid of a hypothetical tenant can influence the rateable value. Therefore factors which reduce the life-expectancy of a building and thereby affect the capital value of a property but not its annual value can be ignored.

5.3.12 In *Lloyd (VO) v. Rossleigh, Ltd.* (1962), the Court of Appeal held that the rateable value of a showroom was not affected by the fact that the area was to be a comprehensive development area, because no such proposal had been confirmed. It was therefore considered that a hypothetical tenant would ignore such a proposal in fixing a rent to be paid.

5.3.13 Thus, a distinction must be made between future events which are certain and those which are merely proposed and may not, in fact, occur.

5.3.14 Similarly, factors which influence an actual landlord and tenant relationship but which are not envisaged by the statutory definition of rateable value and the implied hypothetical tenancy can be ignored.

5.3.15 In *Dawkins (VO) v. Ash Brothers and Heaton Ltd.* (1969), the House of Lords reduced the assessment of a hereditament because of the probability that, within a year of the date of the proposal, part of the premises would be demolished for a road-widening scheme.

5.3.16 Thus, where an external agency is expected to bring the occupation to an end in less than a year and such a prospect affects the value of the occupation, then a valuer can consider the extent to which the rateable value is reduced by the actual prospect in the valuation (if at all).

5.3.17 This would not be true if the landlord intended to demolish the hereditament for his own reasons, because the hypothetical landlord is the one involved in the hypothetical tenancy and is deemed to be offering the hereditament for occupation for at least a period of twelve months. In such a case, the prospect of demolition is ignored in fixing the rateable value. Thus, the rateable value could not be reduced by such a prospect, but a full year's rates would not be paid if occupation did not in fact last twelve months (*Burley (VO) v. A & W Birch Ltd.* (1959)).

5.3.18 Notice the distinction between what can be taken into account within the assessment of rateable value, e.g., the effects of influences which are external to the hypothetical tenancy on the rental bid of a hypothetical tenant; and those which must be ignored, e.g., the no-less real effects of the intentions of the actual landlord or tenant, who do not exist within the hypothetical tenancy.

5.3.19 In normal cases, any question as to the hypothetical tenant's ability to pay a rent must be disregarded. But where the demand for a particular hereditament is very restricted, and especially where there is only one hypothetical tenant, ability to pay is taken into account (*Tomlinson (VO) v. Plymouth Argyle Football Co. Ltd.* (1960)). This point has been applied to the assessments of football grounds, cricket clubs, and certain Oxford colleges.

5.3.20 Tenants' rates and taxes do not include land drainage rates or special rates. However, a 'water rate', as a charge for water supplied, is payable by a tenant.

5.3.21 Under the definition of rateable value, the tenant bears the cost of repairs, insurance and the other expenses necessary to maintain the hereditament in a state to command that rent. The obligation to repair in the statutory definition includes internal decorations but it seems the obligation to put the premises into repair if that is necessary prior to occupation rests with the hypothetical landlord (see *Wexler v. Playle (VO)* (1960)) (see 5.4.25).

5.3.22 'Other expenses necessary to maintain the hereditament' include a renewal fund if the occupation is likely to be permanent, e.g., of a statutory undertaking, and if no liability for repairs is included in the actual rent.

5.3.23 Drainage rates, sea defence rates, fishing rates and similar charges on the owner are 'expenses necessary to maintain the hereditament in a state to command that rent'. But such expenses can include expenditure on other land if it is necessary to preserve the physical existence of the hereditament.

5.3.24 The definition of rateable value places the burden of outgoings on the hypothetical tenant. In this respect, rating assessments are based on similar terms to the rents payable under full repairing and insuring leases negotiated in the open market (see 6.4). Thus, the rateable value is the net rent payable by the hypothetical tenant, as at the date of valuation (see 5.4.38).

5.3.25 The Secretary of State has the power (Sch. 6 para. 6) to make regulations prescribing other specific assumptions in certain cases, and can determine that the normal definition of rateable value will not apply to particular cases of hereditament. The regulations which have been made to date relate to hereditaments occupied by gas, water, docks and harbours, electricity supply and railway undertakers (see 6.3).

5.3.26 **Composite hereditaments**
The definition of rateable value for composite hereditaments, none of which is exempt rates, is contained in Sch. 6 para. 2 (1A) 1988 Act, as follows:

... an amount equal to the rent which, assuming such a letting of the hereditament as is required to be assumed for the purposes of ... [a rateable value for a non-domestic hereditament] would reasonably be attributable to the non-domestic use of property.

5.3.27   Thus, for composite hereditaments, the whole of the property is to be valued but the rateable value is the rent which the hypothetical tenant would pay for the non-domestic part only. Notice that the domestic part is not ignored, but its value does not increase the rental value applied to the rateable hereditament.

5.3.28   The composite hereditament is to be valued vacant and to let (see 5.4.6). So, it is not the actual use of the property but the use to which the hypothetical tenant would put the property which must be considered.

5.3.29   The actual occupation provides strong (but not conclusive) evidence of the likely division between domestic and non-domestic property, as will the structural divisions, etc., within the property.

5.3.30   Having decided which part of the property would be used for non-domestic purposes by the hypothetical tenant, the valuer must produce a rateable value for that non-domestic part with the benefit or burden of the living accommodation but without valuing that living accommodation.

5.3.31   This is achieved by either:
(a)   valuing the non-domestic part of the property and then considering what addition or deduction should be made to reflect the advantage or disadvantage of the presence of the living accommodation; or
(b)   valuing the entire property and deducting the value of the domestic part.

5.3.32   For example, an office-block with a caretaker's flat could be valued by multiplying the area available for office use at a price per square metre derived from comparable office premises. The valuer could then establish whether the hypothetical tenant would increase the rental bid because of the presence of a resident caretaker. Alternatively, a value for the whole block, including the caretaker's flat, can be produced and reduced by the rental value of the flat.

5.3.33   The method adopted depends on the nature of the composite hereditament and the comparable rental evidence available to the valuer (see Chapter 6).

5.3.34 *Partially exempt hereditament*

The rateable value of a non-domestic hereditament which is partially exempt from rates is (Sch. 6 para. 2 (1B) 1988 Act):

... an amount equal to the rent which, assuming such a letting of the hereditament as is required to be assumed for the purposes of ... [a rateable value for a non-domestic hereditament] would, as regards the part of the hereditament which is not exempt from local non-domestic rating, be reasonably attributable to the non-domestic use of the property.

5.3.35 Thus, for a property, part of which is exempt rates, the rent of the rateable part is assessed to include such parts of the non-rateable part as are reasonably attributable.

5.4 Principles of assessment

The principles of valuation vary with the type of property involved and the circumstances relating to each case. But there are certain fundamental principles of valuation which apply, regardless of the property type, or the method of valuation adopted.

In addition, the Secretary of State has the power to prescribe 'principles' and 'rules' to be applied when arriving at a rateable value (Sch. 6 para. 2 (9) and para. 3 (1) 1988 Act).

5.4.1 The process of arriving at an annual value was described by Lord Denning MR, (in *R. v. Paddington (VO) ex parte Peachey Property Corporation Ltd.* (1965) at p. 199), as follows:

The rent prescribed by the statute is a *hypothetical* rent ... which an imaginary tenant might be reasonably expected to pay to an imaginary landlord for a tenancy of *this* [hereditament] in *this* locality, on the hypothesis that both are reasonable people, the landlord not being extortionate, the tenant not being under pressure, the [hereditament] being vacant and to let, not subject to any control, ... the period not too short nor yet too long, simply from year to year.

5.4.2 Case law has established the following principles which should be used, where applicable, to assess the correct rateable value for the hereditament.

5.4.3 *Individual assessment*

As with any valuation, each property must be valued separately and independently (*Ladies Hosiery and Underwear Ltd. v. West Middlesex Assessment Committee* (1932)).

5.4.4   As a result, where all the hereditaments in a rating list were increased by 25% uniformly, it was held that the list was bad (*Stirk & Sons, Ltd. v. Halifax Assessment Committee* (1922)).

5.4.5   Similarly, if a non-statutory formula is used to value a hereditament, it will not be relied on in valuation proceedings without an individual valuation of that hereditament (*Dawkins (VO) v. Royal Leamington Spa Corporation & Warwickshire County Council* (1961)).

5.4.6   *Vacant and to let*
To consider how much rent a hypothetical tenant would pay for a property implies the prospect of occupation, which itself implies that, at the moment of rent being agreed, the property is vacant and, in view of the fact that negotiations are going on, to let.

5.4.7   Although the property may, in fact, be occupied, it must be assumed to be vacant and to let in order to ascertain a rateable value (*London County Council v. Erith & West Ham* (1893)).

5.4.8   *Statutory restrictions*
If the hereditament is assumed to be vacant and available for letting on the statutory terms, it follows that restrictive covenants and other private arrangements affecting the actual tenancy of the hereditament are irrelevant when ascertaining its value for rating purposes under a hypothetical tenancy (*Byrne v. Parker (VO)* (1980)).

5.4.9   On the other hand, every potential hypothetical tenant must be affected by statutory powers or obligations attaching to a hereditament regardless of its ownership (ibid.). Thus, any obligation imposed from outside the actual tenancy, e.g., compulsory purchase, will affect the reasoning of the hypothetical tenant. It is a matter of valuation as to whether it will affect his rental bid as at the date of valuation.

5.4.10   An exception to this rule is that statutory restrictions affecting the rent paid for a hereditament are ignored as a result of the decision of the House of Lords in *Poplar Assessment Committee v. Roberts* (1922). In that case, it was held that the annual rent which a hypothetical tenant might reasonably be expected to pay for a hereditament may be something more than can be recovered from any tenant under legislation.

5.4.11   In relation to the pre-1990 rating of dwelling-houses, it has been held, for example (in *O'Mere v. Burley (VO)* (1968)), that a 'fair rent' determined by the Rent Officer under the Rent Acts has no direct bearing on the value of a hereditament for rating purposes, because the statutory definition of a 'fair rent' is not the same as that of a rateable value.

5.4.12 ***Rebus sic stantibus***
The actual property is to be valued as it in fact is – *rebus sic stantibus*, i.e., things as they stand.

5.4.13 The property must be valued as it exists as at the date of valuation, with all the then existing circumstances, and not as it once was, or as it might become in the future (*Robinson Brothers (Brewers) Ltd. v. Houghton and Chester-le-Street County Assessment Committee* (1938)).

5.4.14 The principle of *rebus sic stantibus* is considered particularly in relation to three aspects of the property:

5.4.15 *Structural Alterations*
The principle of *rebus sic stantibus* means that no structural changes may be envisaged when valuing a hereditament. While premises are undergoing alteration, they must be valued in the state in which they are at that time. If structural alterations are likely in the future, then the hereditament must be valued as it is now and revalued in the future when the alterations are complete.

5.4.16 But the making of a minor alteration of a non-structural nature, provided that it is not so substantial as to change the mode or category of user may be taken into account (*Re Appeal of Sheppard (VO)* (1967)).

5.4.17 *Mode of Use*
Property should normally be valued for the use to which it is put at the date of the proposal.

A dwelling-house must be assessed as a dwelling-house; a shop as a shop but not as any particular kind of shop; a factory as a factory, but not as any particular kind of factory. (*Fir Mill Ltd. v. Royton U.D.C. & Jones (VO)* (1960) (at p. 185)

5.4.18 Although the property has to be valued vacant and to let, it is a generally-held principle of rating that a rental value is to be fixed on the basis of the mode or category of occupation of the actual occupier (*Midland Bank, Ltd. v. Lanham (VO)* (1978).

5.4.19 It is recognised that any limitations imposed by planning law may restrict the range of potential hypothetical tenants by limiting the use of the hereditament more narrowly than the *rebus sic stantibus* rule (*London Transport Executive v. London Borough of Croydon & Phillips (VO)* (1974)).

5.4.20 Although *rebus sic stantibus* involves valuing property for the purpose for which it is used at the time, it may be possible (or even necessary) to value a hereditament for a purpose for which it is not currently being used. This situation may occur, for example, where there is no rental evidence available for the current use; where the property is not being used for its most valuable use; or where a property is not being put to any use at all.

5.4.21 In such a situation, a valuer can include all alternative uses to which the hereditament, in its existing state, could be put in the real world, which would be in the minds of competing bidders in the market, where such competition could be established by market evidence, and considering those uses within the same mode or category (*Midland Bank, Ltd. v. Lanham (VO)* (1978)).

5.4.22 Most properties will cause few problems, with the mode of occupation being quite obvious.

5.4.23 In each case, it must be envisaged that no structural alterations are permitted (see 5.4.15–16); that any planning permission required must be available for a potential hypothetical tenant; and, where the property is used, that the mode of occupation is unchanged.

5.4.24 *State of repair*
Traditionally, this is an exception to the *rebus sic stantibus* rule because of the repair liability expressed in the statutory definition of rateable value.

5.4.25 Thus, it has been established that the hypothetical tenant (who is responsible for all items of repair, etc., in the definition of rateable value (see 5.3)) must be assumed to have fulfilled that repair covenant and that, therefore, the hereditament is in a reasonable state of repair. This allows a valuer to ignore the actual state of repair because assuming that the property is occupied by a hypothetical tenant means assuming that the repair covenant has been observed (*Causer v. Thomas (VO)* (1957) and *Wexler v. Playle (VO)* (1960)).

5.4.26 Case law has established that a standard of repair is to be assumed, commensurate with the type of property, regardless of its actual state of repair. In *Calthorpe v. McOscar* (1923), the standard of maintenance was recognised (at p. 275–6) as being the lower of either:

> ... (a) ... such condition as I should have expected to find them had they been managed by a reasonably minded owner, having full regard to the age of the buildings, the locality, the class of tenant likely to occupy them, and the maintenance of the property in such a way that only an average amount of annual repair would be necessary in the future or (b) in such state of repair as would satisfy the requirements of reasonably minded persons who would be prepared to take on lease the [hereditament] ... and on such conditions as to rent as would presume the premises being put at the commencement of the term free of expense to the lessee in such a state of good and sufficient repair as would render only an average amount of annual expenditure necessary ...

5.4.27   Although ordinary lack of repair may not be taken into account, serious defects, particularly structural ones, may be taken into account in certain circumstances. However, it is permissible to take into account when assessing rateable value structural defects which it would be unreasonable or impossible for a landlord to repair (*Saunders v. Maltby (VO)* (1976)).

5.4.28   Thus, the hypothetical tenant is assumed to be able to negotiate a reduced rent only if, in fact, the state of disrepair is such that a reasonably-minded landlord would not put the hereditament into good repair (ibid.). This, of course, implies that the responsibility for putting the property into good repair at the beginning of the term rests with the landlord and that the tenant takes the property in good repair and undertakes only to maintain that level of repair.

5.4.29   **Tenant fresh on the scene**
The hypothetical tenant is assumed to inspect the hereditament at the date of valuation and, having made reasonable inquiries as to the property, locality, etc., to make a rental bid at that time.

5.4.30   It is, therefore, not possible for the hypothetical tenant to compare the current conditions affecting the property with those which existed previously (*Finnis v. Priest (VO)* (1959)).

5.4.31   What must be determined is the effect on the mind of the 'reasonably-minded person' who has no previous knowledge of the property or area, of conditions as at the date of valuation.

5.4.32   **Consider all evidence**
As with valuations for any other purpose, all factors which affect the rental value of property (other than those which rating law specifically

excludes) must be taken into account in assessing rateable value (*Staley v. Castleton* (1864), *Beath v. Poole (VO)* (1973)).

5.4.33   The extent of the evidence to be treated as available should be considered on the following principles – rateable value is the rent which a tenant might reasonably be expected to pay, and in estimating this rent all that could reasonably affect the mind of the hypothetical tenant ought to be considered (*Robinson Brothers (Brewers) Ltd v. Houghton and Chester-le-Street Assessment Committee* (1937)).

5.4.34   The future may be more or less uncertain and this affects the *weight* given to such evidence. It does not render any evidence of expectation inadmissible.

5.4.35   It is important to distinguish inadmissible evidence from evidence to which little or no weight is attached. Inadmissible evidence cannot be considered by the valuer or the courts and should not even be mentioned in a court case.

5.4.36   Admissible evidence should be considered by the court and if the court considers that the evidence is not strong, or not persuasive, or that there is better evidence available, then the court will not rely heavily on such evidence. But it can at least be presented and relied on to some degree. This is not true of inadmissible evidence (see *Robinson Brothers (Brewers) Ltd. v. Houghton and Chester-le-Street Assessment Committee* (1937) and *Garton v. Hunter (VO)* (1969)).

5.4.37   The Lands Tribunal has adopted the rule in rating cases that events, including rents fixed after the date of valuation, are only admissible as evidence of value in order to prove or disprove a trend or anticipation established at the date of valuation.

5.4.38   *Valuation date*
As with all valuations, it is essential to have a date at which to value the property. However, in rating this may be a complicated issue.

5.4.39   Briefly, the hereditament is valued in its physical state as at 1 April 1995, being the date the list took effect (or, if the list is being altered, as at the date of the proposal or the date when the list is altered (see Chapter 9 Appeals)) but with values as they existed on 1 April 1993 – the antecedent valuation date (AVD). For the list which will take effect on 1 April 2000, the AVD will be April 1998.

5.4.40   Thus, under Sch. 6 para. 2 (3) of the 1988 Act, the date of valuation for the rating list which took effect on 1 April 1995 is as follows:
(a)   1 April 1993 is the date by reference to which the level of values and 'non-mentioned matters' are to be taken into account (the antecedent valuation date). (Non-mentioned matters are anything which is not 'mentioned matters' (see 5.4.41));
(b)   1 April 1995 is the date by reference to which 'mentioned matters' are to be taken into account, when valuing for the preparation of the list. ('Mentioned matters' are defined in 5.4.41));
(c)   the date of the proposal or the date the list is altered is the date by reference to which 'mentioned matters' are to be taken into account, when valuing for an alteration to that list.

5.4.41   'Mentioned matters' are defined (ibid.) as:
(a)   matters affecting the physical state or physical enjoyment of the hereditament;
(b)   the mode or category of occupation of the hereditament;
(c)   matters affecting the physical state of the locality in which the hereditament is situated or which, though not affecting the physical state of the locality, are nonetheless physically manifest there; and
(d)   the use or occupation of other premises situated in the locality of the hereditament.

5.4.42   Thus, a hereditament is valued for a new rating list (see Chapter 8) in its condition as at the date the list takes effect, i.e., 1 April 1995, but at values existing at the antecedent valuation date, i.e. 1 April 1993, for the current list.

5.4.43   Where an entry relating to a hereditament is altered during the life of a list (see Chapter 9), the hereditament is valued in its physical condition as at the date a proposal is made to alter the list, but at values existing at 1 April 1993.

5.4.44   Values existing as at 1 April 1993 should, therefore, provide the level of values which apply to all rateable values for as long as the rating list remains in force. This level of values may be called the 'tone of the list' and is the value established as at the antecedent valuation date, which is used to value all hereditaments regardless of how much they alter physically during the life of the list (see Chapter 6 and Chapter 7).

5.4.45   The valuation date should be considered within an historical context in order to be fully appreciated (see Appendix D).

5.5   Check-list

5.5.1   Rateable value is the annual rent payable for the hereditament by a hypothetical tenant who is responsible for repairs and all other outgoings (5.3.1–5.3.2).

5.5.2   Rateable value assumes a hypothetical tenancy with a hypothetical tenant who pays the rent (5.3.3–5.3.24).

5.5.3   Each hereditament must be valued individually (5.4.3–5.4.5).

5.5.4   Hereditaments are assumed to be vacant and to let (5.4.6–5.4.7).

5.5.5   Statutory restrictions (except those relating to the level of rent payable) affect the rateable value, but private restrictive covenants do not (5.4.8–5.4.11).

5.5.6   The hereditament must be valued *rebus sic stantibus*, i.e., things as they stand, particularly with reference to structural alterations and mode of use: state of repair is an exception to this rule (5.4.12–5.4.28).

5.5.7   The hypothetical tenant is assumed to be 'fresh on the scene' and not able to compare conditions which used to exist with those which exist at the date of valuation (5.4.29–5.4.31).

5.5.8   All factors which affect the value of hereditaments must be taken into account (5.4.32–5.4.37).

5.5.9   The valuation date is 1 April 1993 for levels of value and for all factors, except the physical state of the hereditament and its locality, the mode of occupation of the hereditament and its locality and any other matters affecting the physical state of the locality, all of which are assessed as at 1 April 1995 (or the date of the proposal, if the hereditament is revalued following the list taking effect) (5.4.38–5.4.45).

Chapter 6

Methods of valuation

6.1   Synopsis

6.1.1   With one major exception, any of the traditional methods of valuation may be used to find a rateable value.

6.1.2   The one exception is that certain hereditaments, often described as those occupied by 'statutory undertakers', have their rateable values fixed in accordance with rules laid down in legislation.

6.1.3   With this exception, the use of rental evidence, profits method and/or contractor's test are acceptable methods of fixing rateable value, provided that they are adapted to conform to the definition of rateable value.

6.1.4   There are certain property types which are normally valued using one of the traditional valuation methods, and while one method may be more appropriate than the others, this does not mean that evidence from other methods of valuation should be ignored. It merely means that more reliance is likely to be placed on the most suitable method.

6.1.5   Regardless of the method of valuation used, it is common for the rating assessments of comparable hereditaments to be used to support other valuation evidence, once the tone of the list has been established.

6.2   Introduction

6.2.1   The Local Government Finance Act 1988 does not prescribe the methods of valuation to be used in determining rateable values.

6.2.2   With the exception of the assessments of the so-called 'statutory undertakers' hereditaments, any of the traditional methods of valuation (i.e., rental evidence, profits method and contractor's test) can be used to fix a rateable value.

6.2.3   In addition, the use of comparable rating assessments is important both to support valuations and to maintain the tone of the list.

6.2.4   Nevertheless, the Secretary of State for the Environment has been given power (paras. 2 (9) and 3 (1) of Sch. 6 1988 Act) to prescribe 'principles' and 'rules' to be applied when arriving at the rateable value of a non-domestic hereditament. Such 'rules' include the imposition of a decapitalisation rate for use within a contractor's test method of valuation (reg. 2A of the Non Domestic Rating (Miscellaneous Provisions) (No. 2) Regulations 1989 (SI 1989 No. 2303) as amended) (see 6.6.23).

6.2.5   Thus, except in those cases where the Secretary of State has laid down 'principles' or 'rules', there is no legal requirement to use any particular method in order to arrive at rateable value.

6.2.6   It has been known for a court to accept the opinion of a qualified valuer, that the rateable value of a particular hereditament was £x without any justification or supporting evidence as to how or why that assessment was produced. But usually a valuer will use one or other of several established methods of valuation.

6.2.7   In *Garton v. Hunter (VO)* (1969), valuations on the contractor's test and on the profits basis were submitted in respect of a caravan camping site. The Court held that these valuations were not excluded by evidence of an open-market rent paid for the site under a recent lease, and that all relevant evidence was to be admitted to the Court: the goodness or badness of it went to its weight as evidence, not its admissibility.

6.2.8   However, the fact that rateable value is a net annual rent (defined in 5.3.1), means that the best way to find a rateable value is to use suitable rental evidence available on the hereditament to be valued (the subject hereditament) or comparable hereditaments (e.g., *Garton v. Hunter (VO)* (1969) (see 6.4).

6.2.9   Although it is necessary, under the statutory definition of rateable value, to find a rent for the hereditament, it must not be assumed that a rent which is actually being paid for the hereditament is conclusive evidence of value. Rents need to be tested to ensure that they are suitable and reliable for the fixing of a rateable value.

6.2.10   Though the rent actually paid may not necessarily be equivalent to the rateable value, or even conclusive evidence of value at the date when the rent was fixed, if a rent payable under a yearly tenancy has been fixed at the valuation date, without payment of any premium, etc., it may be taken as prima facie evidence of value, liable to be disproved (*Poplar Assessment Committee v. Roberts* (1922) and *Garton v. Hunter (VO)* (1969)).

6.2.11   This may be done in many ways. It may be shown, for example, that the rent paid is not an open-market rent fixed at arm's length, i.e., it was agreed between connected parties; or that the rent included some other property, such as goodwill.

6.2.12   Nonetheless, actual rents, where they are available and cannot be challenged, should be the best guide to rateable value. Thus, it is usual for rating valuers to investigate the availability of rental evidence first when valuing a hereditament (see 6.4).

6.2.13   Where the so-called best evidence, i.e., the actual rent payable for the hereditament or the rents of similar comparable hereditaments, is not available, the motive likely to induce potential tenants to bid for the hereditament is a relevant factor in estimating what the amount of such bids might reasonably be expected to be, and therefore in determining the rateable value.

6.2.14   Whilst profits are not rateable, it is recognised that the ability to earn profits may affect the rent which a tenant can afford to pay for premises and, in the absence of more direct evidence of rental value, afford some guide as to the rent which can be expected.

6.2.15   Thus, a profits method of valuation can be applied to property types which are not usually rented but where an element of profit, normally derived from some unique feature (e.g., a licence to trade), is a factor in its use (see 6.5).

6.2.16   Where property is of a kind which is rarely rented and where there is little or no element of profit in the undertaking, it is possible to assess a rental value based on its capital value or on the actual cost of land and buildings, as a guide to fixing the rateable value. In such circumstances, use of the contractor's test is made (see 6.6).

6.2.17   It is, however, important to recognise the distinction between cost and value (see 6.6.28). Nevertheless, the contractor's test is a recognised method of valuation for certain property types.

6.2.18   Although hardly a method of valuation, the rateable values of comparable hereditaments are admissible as evidence of value, either as a means of arriving at a rateable value or to support the results of the other methods of valuation (*Pointer v. Norwich Assessment Committee* (1922) and *Shrewsbury Schools v. Shrewsbury Borough Council and Plumpton (VO)* (1960) and *J. Sainsbury Ltd. v. Wood (VO)* (1980)).

6.2.19   This tends to be the case only once a new rating list has been established for some time (generally after one year) and where the level of value or tone of the list (see 5.4.44) has been confirmed either by the passage of time or by valuation tribunal and/or Lands Tribunal decisions.

6.2.20   However, an assessment under appeal is customarily excluded from consideration as being unlikely to provide reliable evidence of value (*Thomas Scott & Sons (Bakers) Ltd. v. Davies (VO)* (1969)) and, with this exception, a valuation officer cannot deny the correctness of an assessment of a hereditament in his own area, whether made by himself or a predecessor.

6.2.21   Entries in the rating list are, therefore, statements of value by the valuation officer.

6.2.22   RICS/ISVA referencing guidelines are adopted, unless local customs vary these (see RICS/ISVA (1993) *Code of Measuring Practice* The Royal Institution of Chartered Surveyors and the Incorporated Society of Valuers and Auctioneers).

6.3   Assessment by statutory formula

6.3.1   Since the 1930s, the contribution to local rates made by a number of undertakings formerly assessed on the profits basis has been laid down by Parliament. Many such undertakings were not profit-making and therefore the application of the profits basis would produce an unsatisfactory assessment.

6.3.2   Such so-called statutory undertakers have, therefore, had their rating assessments fixed by legislation – either by the rateable value being declared in statutory instruments or by the method of calculating the rateable value being set out in statutory instruments.

6.3.3   Since privatisation of such industries as electricity and gas, the status of such undertakings has reverted to profit-making and central government has declared its intention to re-establish the suitability of a profits method for such undertakers by the next revaluation on 1 April 2000. To date, the only such legislation to have been passed is the British Waterways Board and the Telecommunications Industry (Rateable Values) Revocation Order 1994 (SI 1994 No. 3281), which, with effect from 1 April 1995 frees the rateable values of such hereditaments from statutory controls.

6.3.4   A committee has been established to investigate the treatment of plant and machinery involved in such operations following the proposed legislative changes.

6.3.5   *Electricity premises (see 8.4.11)*
The Electricity Supply Industry (Rateable Values) Order 1994 (SI 1994 No. 3282) splits electricity hereditaments into those which appear in local lists (see 8.3) and those which appear in central lists (see 8.4).

6.3.6   Certain hereditaments are entered into local lists. Under article 5, for example, hereditaments comprising land, plant or buildings used or available for use for the purpose of generating electricity, where its primary source of energy is wind, tidal, water power or the burning of refuse, are entered into local lists. The rateable value of such hereditaments is fixed at a given value per megawatt of the declared net capacity of the generating plant (article 6).

6.3.7   Those electricity hereditaments which are entered into the central list are identified by purpose and by occupier. For electricity generation, occupiers such as National Power plc and Powergen plc are listed; for electricity transmission the occupier is the National Grid Company plc; and for electricity distribution, occupiers include South Wales Electricity plc, Midlands Electricity plc and Manweb plc. The rateable values of such hereditaments are stated, with effect from 1 April 1995, together with recalculation factors.

6.3.8   *Gas premises (see 8.4.12)*
The British Gas plc (Rateable Values) Order 1994 (SI 1994 No. 3283), provides a standard formula for the determination of rateable values for operational gas hereditaments in England and Wales, based on a given rateable value for the year beginning 1 April 1995.

6.3.9   In subsequent years, this amount is recalculated according to a standard formula, in accordance with article 6, which takes account of changes in the estimated length of gas pipelines.

6.3.10   *Railway premises (see 8.4.13)*
The rateable values of railway hereditaments which fall to be entered in the central rating list, including a recalculation factor for subsequent years which is based on variations in traffic, are contained in the Railways (Rateable Values) Order 1994 (SI 1994 No. 3284).

6.3.11   *Water supply hereditaments (see 8.4.15)*
The rateable values of water supply hereditaments which fall to be entered on the central rating lists are contained in the Water Undertakers (Rateable Values) Order 1994 (SI No. 3285), which also provides for annual adjustment by reference to variations in water supply.

6.3.12   *Central rating list*
In order to deal with the difficulty of apportioning the value of network industries between billing authority areas, the above hereditaments are entered in the central rating lists (see 8.4) but certain specified electricity hereditaments and statutory dock and harbour undertakings are entered into the appropriate local rating lists (see 6.3.7).

6.3.13   *Other hereditaments*
Certain properties, such as schools and universities, are valued by reference to an 'informal' formula agreed between the valuation officer and the appropriate representative bodies. This formula is not governed by statute and, if presented before a valuation tribunal, must be justified purely on valuation grounds.

6.3.14   Under reg. 3 of the Non Domestic Rating (Miscellaneous Provisions) (No 2) Regulations 1989 (SI 1989 No. 2303), in ascertaining the rental value of a hereditament 'occupied by a public utility undertaking', any relevant evidence is to be taken into account.

6.3.15   Despite the fact that the formulae have been designed to replicate as far as possible the valuations which would be made if other, better evidence existed, the government has stated its intention to return all such properties to conventional methods of valuation.

6.3.16   As with conventionally-assessed hereditaments, the assessment of properties valued by statutory formulae are reviewed to reflect physical changes to the hereditament and its locality.

6.3.17   The process is carried out annually and the method of carrying out the task is also specified in the appropriate Regulations, and is subject to 'tone' provisions. The assessments can be challenged by the appropriate 'interested persons' through the proposal procedure (see Chapter 9) and at valuation tribunal and thence Lands Tribunal (see 10.16).

6.3.18   It seems that it is also possible to challenge the Central Valuation Officer, or the Secretary of State, on the grounds that they have failed to perform their duties properly or fairly by judicial review.

6.4   Rental evidence

6.4.1   The assessment of such hereditaments as shops, offices and certain kinds of industrial premises is usually derived from actual rents.

6.4.2   As stated earlier at 6.2.8, actual rents are considered to be the best evidence on which to base a rateable value, since the definition of rateable value requires the fixing of a rent. Thus, it is logical to assume that suitable rents of comparable hereditaments would provide the most reliable evidence.

6.4.3   However, the mere fact that a rent is passing on a property is not sufficient for that rent to be used in fixing the rateable value. That rent must be suitable as a basis for fixing rateable value and be reliable evidence before it can be used.

6.4.4   It is necessary to consider which rents are suitable as evidence for rating purposes and any adjustments necessary in order to compare them with the statutory definition of rateable value, i.e., a net annual rent (see 5.3.1).

6.4.5   *Suitability and reliability*
There are two basic types of rental evidence:
(a)   direct rental evidence, which is the rent paid on the hereditament which is being valued; and
(b)   indirect rental evidence, which is the rent paid on a hereditament similar and comparable to the hereditament which is being valued.

6.4.6   In both cases, the considerations applied to rent passing must be similar. Where indirect evidence is involved, it is essential to ensure that the hereditaments themselves are as similar as possible. It is a basic principle of valuation that the valuer should always compare like with like.

6.4.7   It is possible to look at the actual rent passing on the hereditament to be valued because the actual occupier must be regarded as a possible hypothetical tenant (see 5.3.7). However, the rent which the occupier actually pays is not necessarily the measure of the rateable value.

6.4.8   The definition of rateable value (see 5.3.1) is very close to the rent usually paid in the open market for commercial property, being an open-market net rent. However, very few properties are let on annual tenancies and there may be other ways in which the rents paid under actual leases do not comply with the statutory definition of rateable value. For example, clauses inserted in an actual lease may impose additional liabilities on the actual occupier and these in turn may affect the level of rent actually paid.

6.4.9   However, where the actual rent passing does not conform to the statutory definition of rateable value, it may, nevertheless, be capable of adjustment and, in that adjusted form, useful for rating purposes (see 6.4.19–53).

6.4.10   If there is a quantity of open-market rents of comparable hereditaments, then these rents can be used to value another hereditament for which no direct evidence exists.

6.4.11   It is, of course, essential to establish that the hereditaments are comparable in all material respects if the rental evidence is to be suitable for valuation purposes.

6.4.12   As for direct evidence, it is important that the rent be adjusted, if necessary, to comply with the statutory definition of rateable value (see 6.4.19–53).

6.4.13   It is usual in such cases for a rent, having been suitably adjusted if necessary, to be reduced to a unit of comparison (e.g., a price per square metre, a price per square metre in terms of zone A (see Appendix E for details of zoning), a price per seat, a price per bed, etc.) for valuation purposes.

6.4.14   However, in both cases, it is always necessary to ensure that the rent conforms or can be realistically adjusted to conform to the statutory definition of rateable value and to all other rating principles.

6.4.15   The more adjustments necessary and, therefore, the further away from the statutory terms, the less reliable the rent (*Edwards and Mann v. Hatton* (1866)).

6.4.16   **Unsuitable rent**
Rents which are not suitable for rating purposes include:
(a)   rents subject to statutory control (because these limit the level of rent which a landlord can demand and do not reflect the level of rent which a tenant might offer for a hereditament (*Poplar Assessment Committee v. Roberts* (1922); *O'Mere v. Burley (VO)* (1968));
(b)   rents fixed at some distance from the date of valuation, because such rents are unlikely to bear any relation to the value of the hereditament as at the date of valuation;
(c)   rents on long leases without rent reviews, because such rents will not equate to a rent 'from year to year';
(d)   rents agreed between connected parties, i.e., not at arm's length, because of the presumption that such rents were not fixed in the open market;
(e)   rents fixed by tender (*Leisure UK Ltd. v. Moore (VO)* (1974)) and interim rents fixed under Landlord and Tenant legislation;
(f)   sale and leaseback rents, where a reduced rent may be part-consideration for the sale. In *John Lewis & Co. Ltd. v. Goodwin (VO) and Westminster City Council* (1979), the Tribunal held that a sale and leaseback rent was part of a funding operation. Similarly, turnover rents are unacceptable;
(g)   rents fixed as a result of fraud;
(h)   'rent' which includes the price paid for the goodwill of a business previously carried on there or other additional property;
(I)   rent reflecting restrictions or concessions not envisaged in the statutory definition of rateable value;
(j)   rent paid under a lease which was part of a larger transaction.

6.4.17   Unsuitable rents should be ignored entirely.

6.4.18   It is, therefore, necessary to rely uniquely on open-market rents, fixed at arm's length, for the hereditament or similar hereditaments, at or near the date of valuation, which are or can be adjusted to conform to the statutory definition of rateable value (see 5.3.1 and 6.4.19–53).

6.4.19   **Adjustment of rental evidence**
Of course, relatively few actual rents are paid under leases or agreements based precisely on the statutory terms, and most actual rents must, therefore, be adjusted into those terms before they can be used.

6.4.20 The following rents are normally capable of adjustment and use in rating valuations:
(a) rents where the landlord is responsible for rates, repairs, insurance or other outgoings (see 6.4.23 and 6.4.36);
(b) rents agreed in consideration of a premium or where improvements are made as a condition of the lease, whether paid by the tenant or the landlord (see 6.4.27);
(c) rents with a rent-free period (see 6.4.41);
(d) rents following a surrender and renewal of a lease (see 6.4.43)
(e) rents which include a service charge (see 6.4.44);
(f) rents including an amount for the use of trade fittings, fixtures, plant and machinery, etc. (see 6.4.45);
(g) rents with a typical review pattern or stepped rents (see 6.4.47).

6.4.21 The fewer the adjustments, the more reliable the resulting rental evidence. If a rent is not capable of adjustment, then it should be ignored.

6.4.22 All the circumstances surrounding transactions of comparable hereditaments should be investigated and, if possible, the lease itself should be read.

6.4.23 *Rates*
The definition of rateable value (5.3.1) assumes that the hypothetical tenant is responsible for the payment of rates. If rent paid includes rates, then it must be adjusted to exclude rates.

6.4.24 Assuming that no material change in the rates paid was envisaged when the rent was fixed, then deduct, from the actual rent agreed, the amount of rates payable (rateable value × UBR) as at that time.

6.4.25 It may, however, be more appropriate to apply the 'equation theory' to the matter (see Emeny and Wilks, 1984, pp. 187–90). The equation theory assumes that an occupier has a given amount of money out of which to pay both rent and rates. The occupier will not care how that money is split between rent and rates as long as the overall sum is not exceeded. Thus, if the occupier's rate liability is increased, he will require a corresponding decrease in rent and vice versa.

6.4.26 Thus, assume that the inclusive rent (which conforms to rateable value, except that it includes rates) is £9,800 per annum, and the rate payable is 40p, then call the rent (in terms of rateable value) R. The rates paid are R × 0.40. Thus, R + 0.40R = £9,800 and R = £7,000.

6.4.27   *Premiums and capital expenditure in accordance with the lease*
If capital payments are made as a condition of the lease, then it is assumed that the rent paid under that lease has been reduced because of the capital payment. Payments, such as a premium, money spent on accrued repairs, alterations, etc., must be treated as capitalised rent.

6.4.28   The annual equivalent of such capital payments is added to the rent paid to produce full rental value on the appropriate terms. (The annual equivalent is found by dividing the capital sum by the appropriate Years' Purchase.)

6.4.29   In selecting an appropriate Years' Purchase, dual-rate, untaxed tables are used (*Trevail (VO) v. C & A Modes Ltd* (1967)), i.e., the analysis is considered from the tenant's point of view.

6.4.30   The term of the Years' Purchase will depend on the earliest opportunity to increase the rent paid to full rental value.

6.4.31   Where improvements are carried out as a condition of the lease, the amount by which the improvements increase the rental value of the property (if at all) is irrelevant. What is significant is that the tenant has been required to part with a capital sum (but see *Edma (Jewellers) Ltd. v. Moore (VO)* (1975) for the Lands Tribunal's logic on amortisation).

6.4.32   If a 'reverse premium' is paid by the landlord to the tenant, then the annual equivalent of the 'reverse premium' is deducted from the rent in order to find the full rental value of the property.

6.4.33   *Capital expenditure not as a condition of the lease*
Where money is spent on the property not in accordance with the terms of the lease, it is first necessary to establish whether the expenditure has increased the property's rental value, or whether the expenditure is of benefit only to a tenant personally.

6.4.34   Expenditure which increases rental value can be used to establish the improved rental value, but using the annual equivalent of the expenditure is not likely to prove a satisfactory means of finding the improved rental value, because cost does not equate to value.

6.4.35   The analysis of the rents of comparable properties (comparable in the improved state) is probably the best way of establishing the effect of the expenditure.

6.4.36   ***Repairs, insurance and other expenses***
In the statutory definition of rateable value (5.3.1), the liability for such expenditure is laid down as being the tenant's liability. It is, therefore, essential when calculating rateable value that the rental evidence is adjusted accordingly.

6.4.37   It is usual for valuers to ignore actual costs for repairs, and to deduct 10% of the full rental value (5% for external repairs and 5% for internal repairs, which includes internal decoration), and this practice was accepted in *Trevail (VO) v. C & A Modes Ltd.* (1967).

6.4.38   Sometimes these percentages are varied to take account of the characteristics of the particular property and, in unusual cases where the physical nature of the hereditament justifies it, the actual cost of repair is used.

6.4.39   With regard to insurance, a similar practice exists of estimating the liability, without having regard to the correct and accurate estimation of replacement costs for both buildings and rateable plant and machinery (not land), including site clearance, professional fees, inflation and (in case of partial rebuilding) VAT.

6.4.40   Again, care should be used when 'rule of thumb' measures are adopted, to ensure that inaccuracies are minimised.

6.4.41   ***Rent-free periods***
In such cases, the rent may be adjusted in a similar way to fixed/stepped rents, giving a rental value at the commencement date.

6.4.42   Alternatively, the actual, unadjusted rent may be taken from the date of commencement of the lease as reflecting the value of the hereditament in its then state.

6.4.43   ***Surrender and renewal of leases***
The normal surrender and renewal calculation can be used to calculate full rental value under circumstances in which a lease is surrendered in consideration for a new lease.

6.4.44   ***Service charges, etc.***
Where rent paid includes an amount to cover landlord's services, an adjustment must be made in the rent to exclude that amount (*Bell Property Trust Ltd v. Assessment Committee for the Borough of Hampstead* (1940)).

6.4.45   This also applies to such items as fixtures and fittings and non-rateable plant and machinery included in the rent passing, but only if a realistic assessment of their worth can be made.

6.4.46   *Weekly rent*
Weekly rents must be multiplied by 52 in order to convert them to annual rents (*London County Council v. Wand (VO)* (1957)). There is no justification for using a lower multiplier, to allow for voids.

6.4.47   *Length of lease and dates of reviews*
For rating purposes, the rent to be fixed is the rent 'from year to year'. It follows that annual rents will therefore be the best evidence. However, because annual rents are rare, the courts have been asked to decide on whether three, five, seven or even twenty-one year lease rents are reliable rental evidence.

6.4.48   In the past, the Lands Tribunal has been unwilling to accept that the rent fixed without review for a lease for a term of twenty-one years should be regarded automatically as the rateable value. This rent must be tested, particularly as to the state of the market at the date when the lease began and as to the changes in value which have taken place since (*Woolworth (F.W.) & Co. Ltd v. Moore (VO)* (1978)).

6.4.49   If there is market evidence that a year-to-year tenancy commands a different rent than, for example, a five-year fixed term, then that market evidence should be used.

6.4.50   Obviously, the shorter the term, the more suitable the rent paid, and if the lease includes rent review clauses, then the rent at review may equate to full open-market rental value if the terms of the lease require that it should.

6.4.51   The courts have recognised that a rent initially agreed on a twenty-one year lease with seven-year rent reviews is likely to be higher than the rent which would have been agreed for that property had it been let with three-year rent reviews (*Baker Britt & Co. Ltd. v. Hampsher (VO)* 1976).

6.4.52   *Valuation date*
The valuation date is 1 April 1993 for the rating list which took effect on 1 April 1995 (see 5.4.38 ff.) and evidence up to and including that date which is available to a potential hypothetical tenant is assumed to be available also to the valuer in fixing the rateable value.

6.4.53   Evidence available after that date is inadmissible in court, unless it is used to corroborate a trend already established at or before the valuation date of 1 April 1993.

6.4.54   **Testing**
An apparently unsuitable rent may well prove to be the basis for fixing the rateable value, but it must be tested first. Whether any adjustment should be made for length of lease is a question of fact, and evidence that the rent under a lease for a term of years takes account, say, of an anticipated increase in value, would clearly be relevant.

6.4.55   But despite all these possible adjustments, it remains true to say that actual rents are the starting point for fixing the rateable value for all hereditaments of the kind normally available on lease in the open market.

6.4.56   The best rents are unlikely to differ greatly from the statutory definition of rateable value, and therefore minimal adjustment should be necessary to enable them to be used as evidence. This of course means that actual market evidence should be as near as possible to the definition of rateable value.

6.4.57   All admissible evidence should be considered. Depending on the valuer's judgement of its reliability, the greater weight should be placed on the more reliable evidence *(Garton v. Hunter (VO) (1969))*.

6.4.58   It is essential that the valuer tests or 'stands back and looks' at the evidence and the resulting rateable value to be satisfied that the correct result has been achieved. Any valuation should be supported by rating assessments of comparable hereditaments (see 6.7) and uncertainty should be investigated and any unsatisfactory evidence rejected.

6.4.59   Remember that while rents may be the best starting point, they do not always give the correct rateable value for a variety of reasons. There are other ways of arriving at a rateable value (see 6.5, 6.6 and 6.7).

6.4.60   **Unit of comparison**
Having selected suitable rents and adjusted them so that they conform to rateable value, the rents are normally analysed to produce a unit of comparison, so that the rental value of one property can be compared to the rental value of another comparable property or applied to value another comparable property.

6.4.61   The unit of comparison will depend on the type of property involved. For example, shop property is analysed to a price per square metre in terms of Zone A (see Appendix E); for industrial property a price per square metre is used; for schools it is a price per pupil; for cinemas a price per seat; etc.

6.4.62   The main types of properties and the relevant factors to be taken into account are considered separately in more detail in Chapter 7.

6.5   The profits method of valuation

6.5.1   *Introduction*
Where the best evidence (i.e., rental evidence) is lacking or there is a need to support a valuation based on rental evidence, valuers may consider using the profits method of valuation.

6.5.2   The profits method of valuation is applicable to properties the value of which is wholly or partially dependent on their right or opportunity to trade. While this can be said to be true of all commercial properties to one degree or another, the profits method of valuation is reserved for such commercial properties as hotels, public houses, petrol filling stations and certain leisure properties, where the ability to trade is dependent on the existence of a factual or legal monopoly, usually a licence.

6.5.3   The valuation of such property types tends to be something of a specialism and the advice and experience of other professionals may be necessary at various stages of the process.

6.5.4   *Principle*
Although profits are not rateable, the ability to earn profits may affect the rent and, in the absence of more direct evidence, offer some guide as to rental value. If the object of the tenant most likely to rent the premises is to make a profit, then the amount of that profit is likely to affect the rent which the tenant will be prepared to give.

6.5.5   Where the best evidence, i.e., actual rent or the rents of similar comparable hereditaments is not available, the motive likely to induce potential tenants to bid for the hereditament is a relevant factor in estimating what the amount of such bids might reasonably be expected to be.

6.5.6   It is a mistake to suppose that valuation by rental is a process disassociated from the idea of profit ... the questions whether a hypothetical tenant could be found, and what rent he might reasonably be expected to give if he were found cannot easily be solved, if at all, except by estimating what amount of profit the [business] had yielded in the past and was likely to yield in the future. An intending lessee, whether real or hypothetical, would hesitate to pay a rent which was not based upon these data. (per Lord Watson, *Edinburgh Street Tramways Co. v. Lord Provost, etc. of Edinburgh* (1894) at pp. 475–6)

6.5.7   The valuer must stand in the shoes of a prospective hypothetical tenant with the trading accounts available on the valuation date (1 April 1993 for the current rating list (see 5.4.38 *et seq.*)) and ask: what rental bid would be made for the hereditament for the next twelve months?

6.5.8   Generally, if the profit depends upon the personal skill of the tenant and could be made in other premises, as well as in the premises in question, then the expected amount of the profits will not affect the rent. But if the profits can only be earned in the premises to be rated, and can be earned there by any ordinary tenant, then the anticipated profits will materially affect the rent, although the extent of the influence will depend on negotiation, often called the 'higgling of the market'.

6.5.9   It is the facility to make trade profits afforded by a particular hereditament to a hypothetical tenant which will enhance the value of that hereditament, and not necessarily the profits made by a particular trader.

6.5.10   It seems that where the (hypothetical or actual) landlord has a monopoly, or quasi-monopoly, the profits which the tenant can expect to make will affect the letting value, and that the rent paid will depend on:
(a)   whether the tenant is the only person who can earn the profits likely to be made; and
(b)   whether the hereditament is the only hereditament in which the profits can be made, i.e., whether the landlord's monopoly is complete or not.

6.5.11   If premises are occupied for the sake of making a profit,

any restrictions which the law has imposed upon the profit-earning capacity of the undertaking must of course be considered in estimating the rateable value. (*London County Council v. Erith & West Ham* (1893) at p. 592).

6.5.12   Statute can limit profits in two ways:
(a)   by limiting the charges which a trader can impose between himself and the public; and
(b)   by appropriating the whole or part of earned profits to particular objectives.

6.5.13   Case law shows that limitations described in (a) above must be taken into account, but limitations described in (b) must not.

6.5.14   It is accepted that the hypothetical tenant must occupy under the same statutory restrictions as the actual tenant.

6.5.15   Basically, the profits method of valuation requires an estimation of the average annual gross earnings from the property and the deduction from this of average annual working expenses (excluding rent and rates) and an amount (tenant's share) to reflect the tenant's remuneration, which should include a return on capital invested as well as an element to reflect the risk involved in running the business. The balance represents the amount which the tenant could afford to pay in rent and (because they are based on rent) rates. (see Figure 6.1)

6.5.16   It can be seen that it is the profitability of the occupier's business which is valued in order to assess the rent which the occupier can afford to pay for the premises which accommodate the business.

6.5.17   It is, therefore, necessary to consider the profits method of valuation in relation to such properties as 'properties with trading potential', defined (Colborne and Hall, 1992, p. 43) as:

a property which is designed or adapted for a particular purpose and fully equipped and trading as an operative entity.

6.5.18   As indicated above, the kind of property which is generally valued using the profits method is a property for which a licence is required before it can be used to trade legally. The nature of the licence must be investigated prior to undertaking the valuation because the licence may also impose conditions as to the state of the premises and the facilities provided, as well as on the conduct of the trading activity.

6.5.19   The nature of the licence may vary according to the location of the property. In some instances, the statutes and regulations relating to London are different from those relating to the rest of the UK. In addition, local bye-laws, national policy, etc., may impose local variations on specific trades which may affect the property and therefore the rental value.

Gross receipts

Adjusted if necessary, to show only those receipts relating to the running of the undertaking by a hypothetical tenant

Adjusted gross receipts   £___

LESS: Working Expenses

Adjusted if necessary to show only those receipts relating to the running of the undertaking by a hypothetical tenant including repairs and insurance of the hereditament but excluding rent and rates.

Adjusted working expenses   £___

Divisible balance (or net profit)   £

LESS: Tenant's share comprising:
a. interest on capital invested, including cash in bank, fixtures and fittings, stock, contingency fund, etc.
b. risk
c. remuneration
Tenant's share is usually taken as a percentage of either gross receipts, tenant's capital or divisible balance

Tenant's share   £___

Rent and rates   £___

To calculate rateable value, let $RV = X$

Rent and rates $= X + (X \times UBR)$

Figure 6.1 Profits method of valuation.

6.5.20 Similarly, building regulations, regulations relating to public health, sanitary appliances, fire safety, safety at sports grounds, occupier's liability, general legislation regulating the workplace and employment, Sunday observance, planning and the supply of water for such properties offering public fishing and navigation activities all require full investigation (see Marshall and Williamson, 1997)

6.5.21   *Fixtures and fittings*
Also, because of the requirement to value the property as a trading enterprise, it is necessary to include all tenant's fixtures and fittings, because, while these are not allocated a specific value, not to have them would mean that the levels of income and expenditure could not be achieved and that therefore the profit (out of which the rent is notionally payable) would be impossible to make.

6.5.22   *Goodwill*
Similarly, it is necessary to include within the valuation (again, generally without allocating to it a specific value) the goodwill which attaches to the trade.

6.5.23
Goodwill has been defined as an intangible but marketable asset founded on the probability that customers will resort to the same [business] premises . . .' (Jones Lang Wootton (1989) p. 87)

6.5.24
It is necessary to value both the goodwill which attaches to the property and the goodwill which attaches to the site in the profits method.

6.5.25
It is to be assumed that the trader represents the hypothetical tenant and is of average skill and ability. Where this is not the case (e.g., the trader is demonstrably incompetent in running the business or the trader is a well-known personality, as a result of which trading profits greatly exceed those which would be achievable by anyone else), allowance must be made in the profits method to reflect this.

6.5.26
Remember that, while using the accounts of the actual trader, it is the accounts of a hypothetical tenant which are to be used to fix rateable value.

6.5.27
It may be that the use of the profits method will show no rent available for the landlord. This is acceptable, but in such cases a tenant's overbid should be considered.

6.5.28
An overbid is rent paid to a landlord in excess of the monetary worth to the occupier, and may exist where an occupier has a extraordinary desire or obligation to occupy premises, regardless of market rental values.

6.5.29
An example of such a situation is where property occupied by a local authority is run at a loss and is of a type normally valued by reference to its profit. It may be necessary to increase the 'rent'

produced by the profits method to take into account the additional motives which a local authority may have as a hypothetical tenant over and above those of a purely commercial competitor. This principle of local authority's overbid has been established by the Lands Tribunal in a series of cases concerning municipal markets, a beach and esplanade undertaking and a pier (*Taunton Borough Council v. Sture (VO)* (1958); *Lowestoft Borough Council v. Scaife (VO)* (1960)).

6.5.30   A criticism of the principle of 'overbid' is that the amount of overbid is necessarily an arbitrary figure which cannot be proven by market evidence.

6.5.31   ***Date of Valuation***
Following the decision in *Barrett v. Gravesend Assessment Committee* (1941), as amended by Sch 6 para. 2 (3)(4) of the 1988 Act, the accounts to be considered are those available to the hypothetical tenant at the date of valuation, i.e., the valuation is to be based on the tone of the list (see 5.4.44), and the relevant accounts to be used are the last ones available to the valuation officer when the rating list was prepared.

6.5.32   The practice has been to base the valuation on the last accounts available prior to the date of valuation, but to admit in evidence the latest available accounts to the date of the proposal where these illustrate the working of factors in existence at the date of valuation. Where the latest accounts show the workings of a factor which would have influenced the mind of the hypothetical tenant at the date of valuation, they are admissible as evidence only to show the presence of that influence.

6.5.33   ***Application***
From the decision in *Kingston Union v. Metropolitan Water Board* (1926), the profits method may be summarised as follows.

6.5.34   From the gross receipts of the undertaking for the preceding year, deduct working expenses, an allowance for tenant's profits and the cost of repairs and other statutory deductions and treat the balance remaining as the rateable value plus rates. In outline, the valuation is as shown in Figure 6.1.

6.5.35   Gross receipts and working expenses are ascertained from the actual audited accounts. It is important to remember that the rent to be fixed is the rent which the hypothetical tenant will pay for the hereditament on the statutory terms, and not the rent which the actual occupier would pay.

6.5.36 Ensure that the accounts relate only to the hereditament being valued and do not include other hereditaments or, if they do, that they are adjusted accordingly. Audited accounts are prepared for an individual's (person's or company's) tax purposes. They are not prepared for valuation purposes and may need a large degree of adjustment (perhaps even explanation) before they are suitable for use in a valuation.

6.5.37 Usually three years' audited accounts are required to identify business trends, etc., on which to base the rental value.

6.5.38 Remember that the hypothetical tenant is a trader of average competence. If the actual occupier is above or below average competence, the accounts will need to be adjusted before they can be used in the valuation.

6.5.39 In calculating rateable value, omit from the accounts only repair items which involve capital expenditure, since a tenant would not be expected to carry out such work. It is important to arrive at an average cost for repairs, and it is therefore necessary to consider several years' accounts, in order to achieve a reasonable estimate of annual outgoings and to allow for regular but not always annual items of expenditure. Allowance should also be made for future liabilities, inflation and VAT.

6.5.40 Make necessary additions for sinking funds to replace such depreciating items as buildings, rateable plant and machinery, etc., but not, of course, land. Calculate the replacement cost of each item, and spread it over its predictable useful life.

6.5.41 In assessing the proper share of net profit for the tenant, it is important to allow an appropriate return on capital invested, as well as an element to cover risk, i.e., an inducement to undertake the business and remuneration in running the business. These three items must be included and a percentage of either gross receipts, tenant's capital or divisible balance may be taken to cover them. In all three cases, whichever one is chosen, the higher the risk, the higher the percentage that the tenant will require.

6.5.42 Accounts available after the date of valuation are inadmissible as evidence, unless they are confined to proving trends, etc., indicated in the earlier accounts.

6.5.43 Strictest confidentiality is required when dealing with accounts and no disclosure can be made by the valuation officer to a third party without consent.

6.6 The contractor's test

6.6.1 *Introduction*
Where property is of a kind which is rarely let, interest on the capital value or on the actual cost of land and buildings may be used as a guide to fixing a rateable value.

6.6.2 The measure of rateable value is defined by statute as the rent which might reasonably be expected: interest on cost or on capital value cannot be substituted for the net rental value, but, in the absence of the best evidence (i.e. rents), it can be looked at as *prima facie* evidence in order to answer the question of fact – what rent might a tenant reasonably be expected to pay?

6.6.3 There is judicial authority over a long period of time to support this practice. In fact, even where rental evidence is available, interest on cost or capital value is admissible, although consideration must be given to its weight as evidence (*Garton v. Hunter (VO)* (1969)).

6.6.4 In *R. v. School Board for London* (1885), Cave J, said (as reported in *Ryde on Rating*):

> Interest on cost is a rough test undoubtedly. It is a test in some cases but it is not a test in others. If the place is occupied by a tenant, it is not a good test at all, because the rent which he actually pays is a far better one. If the place is unlet, it is not at all a good test; because it may be that no tenant would give anything approaching to the interest on the cost. But if the place is occupied by the owner himself, then it is in some sense a test, a rough test no doubt, and only prima facie evidence, but still some evidence, to show what the value of the occupation is.... If he could get a place cheaper, at a less rent than the interest on the cost comes to, it is to be assumed he would not go to the expense of building, he would prefer to take the cheaper course and pay the rent.

6.6.5 The contractor's test has been applied to schools (*R. v. School Board for London* (1886)); sewers (*London County Council v. Erith & West Ham* (1893)); a lighthouse (*Lancaster Port Commissioner v. Barrow-in-Furness Overseers* (1897)); a reservoir (*Liverpool Corporation v. Llanfyllin Assessment Committee* (1899)); a farm equipped with sewage plant (*Davies v. Seisdon Union* (1908)), and municipal property, such as town hall (*Chandler (VO) v. East Suffolk County Council* (1958)); a fire station (*North Riding of Yorkshire County Council. v. Bell (VO)* (1958)); a swimming pool (*Woking Urban District Council v.*

Baker (VO) (1959)); a teachers' training college (*Cardiff Corporation v. Williams (VO)* (1973)); an airport (*Coppin (VO) v. East Midlands Airport Joint Committee* (1970)); college and university buildings (*Oxford University v. Oxford Corporation* (1902) etc.); museums (*Cambridge University v. Cambridge Union* (1905)); an old people's home (*Davey (VO) v. God's Port Housing Society Ltd,* (1958)); plant and machinery (*Shell–Mex & B.P. Ltd. v. James (VO)* (1960)); football stadia (*Tomlinson (VO) v. Plymouth Argyle Football Co. Ltd.* (1960)); a cricket ground (*Warwickshire County Cricket Club v. Rennick (VO)* (1959)); and industrial premises (*Cardiff Rating Authority & Cardiff Assessment Committee v. Guest Keen Baldwin's Iron and Steel Co. Ltd.* (1949)).

6.6.6   Although as a result of *Garton v. Hunter (VO)* (1969), cost or capital value can be referred to even where rental evidence is available, it is rare that it will be preferred to actual rents or to evidence based on receipts and expenditure or to a direct comparison with assessments of comparable hereditaments. The contractor's test tends to be a method of last resort.

6.6.7   The Lands Tribunal has frequently rejected the contractor's test in favour of one of the other methods of valuation currently used in rating.

6.6.8   In *Cardiff Rating Authority & Cardiff Assessment Committee v. Guest Keen Baldwin's Iron and Steel Co. Ltd.* (1949), Denning LJ, said (p. 394):

> It is only one of the ways in which the tribunal of fact can form some idea of what a hypothetical tenant would pay.... Even when the contractor's basis is taken, the assessment on that basis is open to great variations up and down, as, for instance, in assessing the effective capital value and in deciding what percentage to take on it.... The possible variations may become so great that the contractor's basis ceases to be a significant factor in the assessment. In such a situation the tribunal of fact may prefer to take some other basis ...

6.6.9   The contractor's test may also be useless as a guide to annual value if the evidence shows that the hypothetical tenant would not have paid a rent based on cost.

6.6.10 *Principle*
What has to be determined is not what the landlord would ask, but what a tenant would give. The landlord is assumed to be a person willing to let to a tenant willing to take, and not a person in a position to dictate terms (*R. v. Paddington (VO) ex parte Peachey*

Property Corporation Ltd. (1965)), so that consideration of the landlord's costs is wrong. However, reference to cost of capital value can be justified from a tenant's point of view.

6.6.11   In *Cardiff Corporation v. Williams (VO)* (1973), Lord Denning MR, (at p. 21) described the following passage (from the Solicitor General in *Dawkins (VO) v. Royal Leamington Spa Corporation and Warwickshire County Council* (1961)) as the 'classic explanation' of the contractor's basis:

> As I understand it, the argument is that the hypothetical tenant has an alternative to leasing the hereditament and paying rent for it; he can build a precisely similar building himself. He could borrow the money, on which he would have to pay interest; or use his own capital on which he would have to forego interest to put up a similar building for his owner-occupation rather than rent it, and he will do that rather than pay what he would regard as an excessive rent – that is, a rent which is greater than the interest he foregoes by using his own capital to build the building himself. The argument is that he will therefore be unwilling to pay more as an annual rent for a hereditament than it would cost him in the way of annual interest on the capital sum necessary to build a similar hereditament. On the other hand, if the annual rent demanded is fixed marginally below what it would cost him in the way of annual interest on the capital sum necessary to build a similar hereditament, it will be in his interest to rent the hereditament rather than build it.

Thus, it can be argued, the contractor's test should produce a ceiling rent above which the rent under the statutory definition should not go.

6.6.12   *Application*
Where cost and value are closely related, as with new buildings, this method of valuation involves little more than the application of appropriate percentages to the cost of constructing the buildings and to the cost of acquiring the land of which the hereditament is to consist.

6.6.13   However, with the rise of building costs and where time and other factors affect the physical state of the hereditament, a wide gap between cost and value means that a modification to the method is essential if it is to produce credible rateable values.

6.6.14   Accordingly, after the cost of construction has been ascertained, it is the practice to reduce the figure to one which represents the 'effective capital value' (ECV) of the hereditament: i.e. the market value of the actual hereditament in a form effective for its purpose at the date of valuation. The appropriate percentage is then applied to the effective capital value.

6.6.15   In *Gilmore (VO) v. Baker-Carr (No. 2)* (1963), the Lands Tribunal recognised five stages in applying the contractor's basis.

6.6.16   *First stage*
Estimate the cost of construction of the building, including any rateable plant and machinery, as at the valuation date, currently 1 April 1993. There is a difference of opinion as to whether it is better to take the cost of replacing the actual building as it is, or the cost of a substitute building providing identical accommodation, but in a modern building. The simple substitute building estimate would seem to be preferred as being the most likely structure which would actually be replaced. Costs include non-remunerative costs, such as exceptional site works and fees, but not VAT.

6.6.17   *Second stage*
Make deductions from the cost of construction to allow for age, obsolescence and any other design and occupational factors necessary to arrive at the 'effective capital value' (ECV). Obsolescence can take various forms, and physical, legal, social, economic and functional obsolescence should all be reflected.

6.6.18   In *Imperial College of Science and Technology v. Ebdon (VO) and Westminster City Council* (1984), the Lands Tribunal considered that an 'effective capital value' could more accurately be described as an 'adjusted replacement cost'. The object of this stage is to adjust the cost of the substitute building to allow for the actual state of the existing building, so that the second stage is not a conversion of cost into value. If there is sufficient evidence of relevant capital transactions, there seems to be no reason, in principle, why the valuation should not begin at this stage.

6.6.19   *Third stage*
Establish the cost of the land as a cleared site with all services available. The principle of *rebus sic stantibus* demands that the land be valued as if limited to its existing use.

6.6.20   The value of the site may reflect advantages or disadvantages which are not accounted for in the ECV of the substitute buildings, but which should be reflected in the adjusted cost (*Imperial College of Science and Technology v. Ebdon (VO) and Westminster City Council* (1984)).

6.6.21   Thus, the deductions made from building costs and land values may be different.

6.6.22   *Fourth stage*
Apply the decapitalisation rate (market rate or rates at which money can be borrowed or invested) to the effective capital value of the buildings and the land. To the 'real rate of interest', add for a borrower's premium, if appropriate, depreciation and the repair liability (*Imperial College of Science and Technology v. Ebdon (VO) and Westminster City Council* (1984) see also 6.6.30–32).

6.6.23   The Secretary of State, under the powers given him to prescribe 'rules' and 'principles' to be applied to valuations, has removed the consideration of an appropriate rate per cent from market consideration. Under the regulations (reg. 2 of the Non Domestic Rating (Miscellaneous Provisions) (No 2) Regulations 1989 (SI 1989 No. 2303), as amended by the Non Domestic Rating (Miscellaneous Provisions) (No. 2) (Amendment) Regulations 1994 (SI 1994 No. 3122), a rate of 3.67% is specified as applying to the notional cost of construction for certain non-profit-making educational hereditaments, hospitals and maternity homes, and a rate of 5.5% in other cases decided under the 1995 rating lists.

6.6.24   The result should be what it would cost the occupier in annual terms to provide the hereditament for himself, rather than to rent it.

6.6.25   *Fifth stage*
The object at this stage is to take into account any items that have not already been considered in valuing the buildings and land, in order to arrive at the annual equivalent of the likely capital cost to the hypothetical tenant.

6.6.26   Such considerations may include the economic health of an industry, but not of an individual company, poor site access and the inflexibility of a district heating scheme (*Imperial College of Science and Technology v. Ebdon (VO) and Westminster City Council* (1984)). However, be careful not to duplicate allowances made at any earlier stages.

6.6.27   *Sixth Stage*
A sixth stage was added by the Lands Tribunal in *Imperial College of Science and Technology v. Ebdon (VO) and Westminster City Council* (1984), and this involves considering whether the result of the fifth stage is likely to be pushed up or down in the negotiations between a hypothetical landlord and a hypothetical tenant, having regard to the relative bargaining strengths of the parties. This could be called the 'stand back and look' stage which is, of course, something every valuer should do for every valuation.

6.6.28 *Value and cost distinguished*

It is very important to distinguish between value and cost. This applies equally, whether the cost is the actual cost of providing the hereditament or the estimated cost of providing a simple substitute building. Two properties may have identical accommodation and be worth the same if they were rented on the open market, but cost different amounts to build. For example, if a property is built on poor soil with expensive foundations at a cost of £100,000 but could have been constructed elsewhere for £75,000, the latter is the basis upon which the calculation must be based. There is also a distinction between expenditure which is unforeseen and accidental, and expenditure which is deliberately undertaken; the former should be ignored and the latter taken into account.

6.6.29

In valuing older buildings, deductions, often large in amount, may be necessary in order to take account of age and the various forms of obsolescence (see 6.6.17). The effective capital value of buildings, some of them built centuries ago, has been determined by visualising either a replacement of them or a substitution for them and then comparing in terms of value the replacement or substitute buildings with the actual buildings. This comparison has usually involved reference to a scale of deductions for age and obsolescence devised and used in the assessment of local-authority schools, either applied strictly or varied according to the judgement of the valuer. It is disabilities which affect, for example, the operating efficiency of the hereditament which should be used to reduce costs to effective capital values.

6.6.30 *Decapitalisation rate*

Before 1990, the percentage of effective capital value which was to be taken as annual value was affected by the motive of the hypothetical tenant in taking the hereditament and the ability of the hypothetical tenant to pay. The application of a market rate to the effective capital value may be inappropriate as that would show what it would cost in annual terms to become an owner of the hereditament. Owners have advantages which occupiers do not have (e.g. a permanent interest in the property, freedom to alter or improve it, the benefit of capital appreciation), so that an annual tenancy is less valuable than ownership, and the annual rent must be fixed below the market interest rate.

6.6.31

The fact that a particular occupier can borrow money at low rates of interest, or can obtain money in the form of benefactories, or government grants, is irrelevant.

6.6.32

The Secretary of State, under the powers given him to prescribe 'rules' and 'principles' to be applied to valuations, has removed the consideration of an appropriate rate per cent from market forces. Under reg. 2 of the Non Domestic Rating (Miscellaneous Provisions) (No 2) Regulations 1989 (SI 1989 No. 2303) (as amended), a rate of 3.67% is specified as applying to the notional cost of construction for certain non-profit-making educational hereditaments, hospitals and maternity homes, and a rate of 5.5% in other cases decided under the 1995 rating lists.

6.6.33 Artificial approach

The Lands Tribunal, while accepting the contractor's basis as 'a poor best', has criticised 'the artificiality of the approach' (*Downing, Newnham, Churchill & King's Colleges, Cambridge v. City of Cambridge and Allsop (VO)* (1968)). However, in *Eton College (Provost and Fellows) v. Lane (VO) and Eton Rural District Council* (1971), the Tribunal said (at p. 182):

Provided a valuer using this approach is sufficiently experienced, and is aware of what he is doing, and knows just how he is using his particular variant of the method, and provided he constantly keeps in mind what he is comparing with what, we are satisfied that the contractor's basis provides a valuation instrument at least as precise as any other approach.

6.7   Comparable rating assessments

6.7.1

Although hardly a method of valuation, reference to rating assessments of comparable hereditaments is widely used in order to ascertain the rateable value of a hereditament where better evidence is lacking, or in order to supplement other evidence.

In *Howarth v. Price (VO)* (1965), the Lands Tribunal said (at p. 199):

Where, however, there is a paucity of satisfactory direct rental evidence, then the best evidence as to rental value is likely to be the 'indirect' evidence provided by the [rateable] values of similar hereditaments; and the greater the similarity between these other hereditaments to the particular hereditament in physical respects such as nature of property, type, age, design, size and surroundings, the better is the evidence so provided. Should there be available as comparables a number of hereditaments all equally similar to the particular hereditament in all material physical respects, then those which are in the same locality as the particular hereditament normally provide better evidence as to rental value than those which are in some other locality, because the former are more likely to be in similar economic sites and therefore the more truly comparable.

6.7.2   The principle was stated by Atkins LJ, in *Pointer v. Norwich Assessment Committee* (1922) (at p. 477):

> In my opinion evidence of the rateable value must be admissible; and for two reasons. In the first place, in cases in which both premises are in the same [authority area], it is evidence against the [valuation officer] in the nature of an admission. And secondly, it may be the only way in which you can get at the rent at which the appellant's premises are worth to let by the year ...

6.7.3   In *Shrewsbury Schools v. Shrewsbury Borough Council and Plumpton (VO)* (1960), the Lands Tribunal re-stated the rule (at p. 322) as follows:

> With regard to the admissibility of the assessments of other public schools, we are definitely of the opinion that such assessments are admissible as evidence on the grounds stated by Atkins LJ, in *Pointer v. Norwich Assessment Committee* [see 6.7.2] ... it seems to us that the admission is no longer effective only in relation to the particular list of the [billing authority] area. The valuation officer is the agent of the Inland Revenue, and in our view, the scope of the admission must now be extended to all lists in England and Wales ... Moreover, the valuation officer in the present appeal, as indeed in all appeals, appears as the agent of the Inland Revenue duly appointed for that purpose, and it seems to us to follow that he is not at liberty to impugn assessments in [rating] lists even though they be outside his own valuation area.

6.7.4   Thus, it would seem that a valuation officer cannot challenge the assessment of a hereditament in his own area, whether made by himself or a predecessor. However, an assessment under appeal is customarily excluded from consideration as being unlikely to provide reliable evidence of value (*Thomas Scott & Son (Bakers) Ltd. v. Davies (VO)* (1969)).

6.7.5   In the production of any rateable value, except on the production of a new rating list, the use of comparable assessments to support better evidence, such as rents, illustrates to the valuation tribunal and to the ratepayer the relative fairness (or otherwise) of the level of rateable value fixed.

6.7.6   The problem of assessing a property type which was not in existence at the date of valuation occurred during the 17 year life of the 1973 valuation list in relation to retail warehouses and out-of-town superstores, amongst others. There was no rental evidence of similar properties in existence as at 1 April 1973 (the valuation date for the 1973 valuation list). The courts resolved the problem of the lack of actual rental evidence by treating rating assessments of similar property types already in the list as the rental value of the property as at 1 April 1973, which could then be used as comparable (rental) evidence in the assessment of new properties or when the assessment of existing properties was challenged (*J. Sainsbury Ltd. v. Wood (VO)* (1980)). Although inherently unsatisfactory, since, of necessity, these assessments had been based on rental evidence of dissimilar shop units, a method of arriving at the rating assessment was achieved.

6.7.7   It has always been possible to use comparable assessments once the rating list has been established and the values confirmed by the valuation tribunal, i.e., when the 'tone of the list' was established (see 5.4.38–45).

6.7.8   When basing the assessment of a hereditament on comparable assessments, the Lands Tribunal (in *Lamb v. Minards (VO)* (1974) said (at p. 162): '... with the passage of time the volume of established assessments acquires weight as evidence of accepted values ...', and it seems that at least twelve months must pass after a revaluation in order to establish (and presumably test) the tone of the list for values of hereditaments in any given area before they can be used as evidence on which to base comparable assessments.

6.7.9   The more unusual the hereditament, the further afield it is possible to go for comparable assessments. For example, in *Shrewsbury Schools v. Shrewsbury Borough Council and Plumpton (VO)* (1960), the Tribunal admitted evidence of the assessments of ninety-nine public schools in all parts of the country. But in the case of ordinary types of hereditaments, such as offices or shops, evidence of assessments outside the locality will usually be of little or no value. However, for all classes of hereditament it is legitimate to refer to assessments throughout the country for the purpose of showing valuation practice and method (*William Hill (Hove) Ltd. v. Burton (VO)* (1958)).

6.8   Check-list

6.8.1   Legislation does not prescribe methods of valuation, except for the so-called statutory undertakers' operational property (6.3).

6.8.2   All methods of valuation are admissible as evidence, the 'goodness or badness' of them goes to their weight as evidence (6.2.7).

6.8.3   Rents are considered to be the best evidence for fixing a rateable value, although they must be suitable and reliable and may need adjustment so that they conform to the statutory definition of rateable value (6.4).

6.8.4   Where there is an element of monopoly, a profits method may be used. While profits are not rateable, they may give an indication of the level of rent which a hypothetical tenant may offer in order to achieve that amount of profit (6.5).

6.8.5   Where a type of hereditament is predominantly owner-occupied and seldom if ever let, recourse may be had to a contractor's test (cost-based approach). While cost does not equate to value, it can be used to identify the potential 'book' rent which a hypothetical tenant would charge himself if he built himself the hereditament (6.6).

6.8.6   In all cases, comparable rating assessments should be used to support rating valuations and may be used as direct comparisons in order to achieve a rateable value (6.7).

Chapter 7

Valuation of usual property types

7.1 Synopsis

7.1.1 All non-domestic hereditaments are valued to rateable value, in accordance with the definition laid down in the Local Government Finance Act 1988 and the principles established by case law (see Chapter 5).

7.1.2 All non-domestic hereditaments are valued using one or a combination of the traditional methods of valuation (see Chapter 6).

7.1.3 All relevant factors which affect the market value of hereditaments are taken into account in fixing the rateable value of non-domestic property, unless rating law and practice states the contrary.

7.2 Introduction

7.2.1 Consideration was given to the basis and principles used to assess rateable value in Chapter 5 and to the traditional methods of valuation in Chapter 6. These two earlier chapters set out the basis of the valuation of hereditaments for rating purposes.

7.2.2 However, in addition to these rules and principles, the nature of the property type to be valued may also affect the valuation methodology.

7.2.3 This Chapter explains very briefly how the general principles of both rating and valuation are applied to specific property types, and any unique features which such properties may possess are highlighted in so far as they affect rating valuations.

7.2.4   It is assumed that the reader is familiar with the basics of valuation techniques, including the RICS/ISVA Code of Measuring Practice, all of which are essential to understanding the practice of valuing property for rating purposes.

7.2.5   The more usual property types (shops, offices and industrial properties) are considered here, together with certain other specialist properties which illustrate how the methods of valuation may be varied to suit circumstances. It should be recognised, however, that in most cases, each is a specialism and that the advice of a specialist valuer may be necessary in practice. This Chapter aims to do no more than indicate features which should be taken into account when valuing these property types for rating purposes. There is no check-list at the end of this Chapter. The reader is referred to the bibliography for further references.

7.3   Shops and other retail properties

7.3.1   Rental evidence is used to produce a rateable value for shops; reference should be made to the methods of selecting, adjusting and testing rental evidence for the purposes of fixing a rateable value described in 6.4.

7.3.2   It is important to recognise that there is a large range of different kinds of shop properties from the small, isolated rural shop, through the parade of suburban shops, the standard units within purpose-built shopping centres, to the (relatively) enormous out-of-town superstores and the city-centre department stores. It is, therefore, extremely misleading to talk of 'the market' for shop properties, when, in fact, each of the shop-property types has a market of its own. These and other principles need to be reflected in rating assessments, as they do in valuations for any other purpose, and the choice of comparable properties must be made in such a light.

7.3.3   Similarly, locality (city centre, sub-urban secondary area, local tertiary (often isolated) or out-of-town centre) is crucial both to the nature of the properties and to the comparables which are useful.

7.3.4   Shop premises are measured to net internal area (or effective floor area), with the sales area (normally ground floor) being of major importance, and the unit of comparison normally being a price per square metres in terms of Zone A (see 7.3.9–18 and Appendix E). Alternatively, the unit of comparison may be a price per square metre overall, with the areas of ancillary accommodation (storage, offices, etc.) given separately. The decision whether to measure an area in terms of Zone A or calculate a price per square metre overall will depend on the nature of the shop property (see 7.3.9–18 and Appendix E).

7.3.5   Shops are normally rented in the open market and this should mean that a large body of rental evidence is available on which to base a rateable value. Rating valuers should, however, be aware of the legislation affecting the rent passing between a landlord and tenant in the open market because, in some situations, the law specifies factors which must be taken into account or ignored in fixing the rent passing. Such factors may not apply in rating law and, therefore, the rent passing may need to be adjusted if it is to be suitable for rateable value purposes.

7.3.6   For example, the Landlord and Tenant Act 1954, which permits a tenant to have a new lease after the expiry of a previous lease, limits the amount of rent payable under the new lease to ensure that any improvements made by the tenant during the previous 21 years are ignored when fixing the new rent. Within the rating hypothesis, where the 'rent' is paid by a hypothetical tenant 'fresh on the scene', such improvements would be reflected in the rateable value. This difference in the basis of 'rent' payable means that the market evidence of rent passing on the property is of little use (without adjustment) in fixing the ratable value of the hereditament. In such a situation it may be that adjustment of the rent is not practicable and that the valuer would find more useful market evidence elsewhere.

7.3.7   Other kinds of rents which may need adjustment or outright rejection include interim rents, rents paid under extensive user restrictions (except those imposed by statute), and rents reduced to attract an 'anchor tenant' (a trader introduced into a trading location to attract the shopping public which also attracts other traders to occupy nearby premises). (See also 6.4 for consideration of rental evidence.)

7.3.8   As with the valuation of shops for any other purpose, the rating valuer should recognise the relative importance of the trading pitches within the retail location. Patterns of value which exist within the market should be reflected within rating valuations, including the influence of prime pitches; anchor tenants; 'dead frontages' (i.e. frontages which do not involve trade and tend to present psychological barriers to pedestrian shoppers and inhibit them from proceeding further, such as hoardings, dwellings and (often) public houses); such important land uses as car parks, which may attract passing trade; and restrictions, such as double yellow lines, which may limit passing trade.

7.3.9   Physical and rental analysis, therefore, will produce patterns of value which must be reflected in rating assessments. Having selected and adjusted rents (see 6.4) to conform to the definition of rateable value, rents can be analysed to a unit of comparison. There are, basically, two methods of analysing shops;
(a)   zoning (presenting retail premises in terms of a price per square metre in terms of Zone A); and
(b)   the overall method (presenting retail premises in terms of a price per square metre overall).

7.3.10   Standard or small shops (i.e. the typical shop in the suburban retail location) are normally analysed using the zoning method of analysis. Zoning is a method of analysing the sales area of a shop which assumes that the front part of a shop in a busy high street is more valuable than the rear, because of its proximity to pavement traffic and the inducement it offers to window-shoppers. The method of zoning varies in different parts of the country; the methodology and objectives are explained in more detail in Appendix E.

7.3.11   For rating purposes, it is arithmetical zoning which is used for rating, with the depth of each zone (except the remainder) known in advance, together with the relative values of any ancillary accommodation.

7.3.12   As with any method of comparison, 'as you devalue, so you value'. Thus, a comparison must be made of 'like with like'.

7.3.13   Ancillaries are measured and often valued in relation to the price per square metre in terms of Zone A. For example, storage accommodation may be valued at one-tenth of Zone A frontage space. Ideally, of course, the rental value of ancillary accommodation should be arrived at by establishing what such accommodation is worth in the open market when separately let.

7.3.14   Consideration should be made of disabilities which affect the rental value of the property, including steps, columns, obsolete layout, inadequate or excessive height, inadequate lighting, and poor visibility from the main passing thoroughfare. The amount of the deductions should be based on market evidence from comparables.

7.3.15   Other allowances include an allowance for quantity (although there must be market evidence to support such an allowance), an allowance for size and an allowance for shape (size along the frontage).

7.3.16   Arithmetical zoning is used by rating valuers and by the Valuation Office Agency, and is accepted by the courts for the valuation of

Table 7.1 Shop premises valued using the zoning method of analysis

		m²	£/m²	£
Ground floor	Zone A	80.0	215.00	17,200
sales	Zone B	47.7	107.50	5,128
	Stock	16.6	26.87	446
Basement	Sales	103.1	21.50	2,217
	Stock	9.7	14.33	139
	Office	2.9	14.33	42
	Staff (under pavement)	16.8	12.29	206
	Stock (under pavement)	21.2	10.75	228
Total				25,606
	Add 10% for return frontage			2,561
				28,167
	Rateable value			£28,150

Based on a valuation presented in *K Shoes Ltd v. Hardy (VO) and Westminster City Council* (1980).

property types to which its principles apply. Table 7.1 shows a typical example of a shop valuation based on zoning.

7.3.17   Zoning is not suitable for 'large' shops, for which it is advisable to use rents analysed to a price per square metre overall. Such property types include out-of-town superstores. Table 7.2 illustrates the valuation of a shop based on a price per square metre overall.

7.3.18   In addition, features to consider within the valuation are likely to include car-parking, vehicular access and automated teller machines (ATMs) (see *Stringer (VO) v. J Sainsbury plc and Others* (1991)).

7.3.19   Despite the obvious unsuitability of the zoning method of analysis for 'large' shops, in the absence of better (rental) evidence of more comparable hereditaments, the zoning method has been recognised by the courts (e.g. in *Trevail (VO) v. C & A Modes Ltd.* (1967)) as being a means by which the rating assessments of larger

Table 7.2 Shop premises valued based on a price per square metre overall

		m²	£/m²	£
Level 3	Main shop	3,317.3	8.00	26,538
	Raised life style	400.9	6.75	2,706
	Life style entrance	14.4	–	–
	Liquor store, cash office and manager's office	88.2	6.00	529
	Warehouse areas	421.8	4.00	1,687
	Loading bay	81.7	3.00	245
	Covered area	196.7	1.00	197
	Plant rooms	71.1	4.00	284
	Plant room 3	11.1	–	–
	Cold rooms	91.6	5.00	458
	Staff rooms, warehouses and offices	74.3	5.00	372
	Store and part plant room	35.0	3.25	114
	Loft (equipment room)	19.3	2.00	39
	Pump room	4.9	2.00	10
	Cleaners' store	2.4	2.00	5
Level 2	Warehouse	326.7	3.00	980
	Stores	35.3	3.00	106
	Canteen and offices	96.1	4.00	384
	Kitchen	38.9	4.00	156
	Air conditioning and plant room	20.0	2.00	40
	Pump room	40.4	1.00	40
	Store under stairs	4.0	2.00	8
				34,898
	Covered parking 191 spaces @ £5 each			955
				35,853
	Less 10% allowance for shape and pillars			3,585
				32,268
	Rateable value			£32,000

Confirmed by the Lands Tribunal in *Wm Morrison Supermarkets plc v. Woffinden (VO)* (1990).

shops can be achieved by an examination of the rents of smaller shops. (See also *J. Sainsbury v. Wood (VO)* (1980).)

7.3.20   The increasing use of turnover rents in centres which are managed by a single property owner gives an alternative method of arriving at a rental value. It should be remembered, however, that in agreeing to receive a rent based on turnover, a landlord's priority would be the success of the shopping centre as a whole and therefore the establishment of a balance of trades within the centre, rather than to optimise the rent payable from each unit or hereditament.

7.3.21   The current trend away from out-of-town retail developments imposed by planning guidance notes, the reduction of car access into city centres, improved security and electronic shopping are all factors which are likely to affect the valuation of retail premises.

7.4   Offices

7.4.1   Offices are valued using rental evidence, reduced to a price per square metre overall of space measured to net internal area or effective floor area.

7.4.2   Where an office block comprises one hereditament, the hereditament will include stairs, passages, etc., and the rateable value will reflect the benefit of all of the accommodation. If such a property is rented, then the rent may be payable under a full repairing and insuring lease and be capable of minimal adjustment before being used as a basis for assessing rateable value.

7.4.3   Where an office block is let in suites to different occupiers, it is likely that the common parts (e.g. stairs and passages) will not be included in any of the hereditaments. Also, it is likely that the tenants of the rented parts pay a rent which reserves to the landlord the responsibility for repairing (and insuring) the structure. Such rents would need to be decreased to reflect the tenant's full repair-liability inherent in rateable value. It would also be necessary to ascertain the nature of any service charge payable for such premises in order to get a full picture of who is responsible for what and at what cost.

7.4.4   The needs of modern office occupiers should be considered when valuing office premises for rating. The effect of inadequate car-parking, IT-based working practices, flexibility of partitions, environmental considerations, and their impact on services provided within office accommodation and any specific occupier requirements (such as the trend for banks to occupy modern, shop-style premises in preference to the traditional fortress-style accommodation) are all factors which should be included within the valuation.

7.4.5   While the choice, analysis and adjustment of rental evidence may be difficult, the valuation of such hereditaments may appear relatively simple. Table 7.3 illustrates a rating valuation for office premises.

Table 7.3 A Valuation of office premises.

Accommodation	m²		
Ground floor	534.1		
First floor	464.5		
Second floor	571.3		
Third floor	473.8		
Total	2,043.7	@ £80 per m²	£163,500
Plus 30 car-parking spaces		@ £100 each	£3,000
Rateable Value			£166,500

Based on the valuation presented in *Tulang Properties Ltd. v. Noble (VO)* (1984).

7.5   Industrial premises

7.5.1   Warehouses and factories are generally measured to net internal area on the basis of a price per square metre overall, together with any office, car-parking and external storage space.

7.5.2   Despite the fact that most industrial premises are used for particular production processes, they should be considered 'vacant and to let' for the potential hypothetical tenant(s) for which they are most suited. The issues of economic and technical obsolescence are likely to be particularly important for this property type.

7.5.3   *Warehouses*
Warehouses are valued using rental evidence and may be part of a larger industrial hereditament (i.e. used together with factory premises) or be separate hereditaments.

7.5.4   The main factors affecting the value of warehouses include accessibility, particularly their proximity to a good road network for distribution purposes and the layout of the building and yard, which should optimise loading and storage operations. The use of fork-lift trucks, for example, requires an absence of columns and changes in floor level within the building, and an optimum eaves' height can be based on the capacity of these trucks to store goods. Tail-board deliveries should ideally be made to covered, raised bays with appropriate levels of security both internally and externally.

7.5.5   There are, of course, older warehouses which do not meet these specifications and their value is reduced as a result.

Table 7.4 A valuation of a warehouse

		m²	% adjustment	£/m²	£
Ground floor	Warehouse	431.1		62.42	26,909
	Offices (original)	16.7	+25	78.02	1,303
	Offices (additional)				
	Front	15.6	+25	78.02	1,217
	Rear including corridor	61.8	+20	74.90	4,629
	Kitchen	2.5	− 5	59.30	148
First mezzanine	Offices	75.2		78.02	5,867
	Front	43.2		10.00	432
	Rear	70.2		–	–
Outside	6 car-parking spaces				900
	@ £150 each				
					£41,405
Rateable value					£41,400

Valuation from *Vincent Bach International Ltd. v. Kubbinga (VO)* (1994).

7.5.6   An illustration of the valuation of a warehouse appears in Table 7.4.

7.5.7   ***Factories***
The value of factory premises varies depending on their use. There are basically two different kinds of factory premises:
(a)   general industrial buildings which can be used by a variety of occupiers; and
(b)   specialised industrial buildings which are purpose-built for a specific kind of occupier.

7.5.8   General industrial buildings are often built within industrial estates and are capable of being used by a variety of occupiers. They are valued using rental evidence.

7.5.9   Once again, factors which affect value include eaves' height, floor loading capacity, accessibility to vehicular/rail transport facilities which are unrestricted by vehicular weight limitations or low bridges, ease of manoeuverability for fork-lift trucks and other moving machinery, loading/unloading facilities, availability of a suitable labour force and public services.

7.5.10   The non-specialist nature of such industrial buildings means that rental evidence can be used to fix a rateable value and therefore any special requirements of a particular occupier can be ignored within the requirements of the hypothetical tenant (see, for example, 5.3.6 and 5.4.7). The valuation of such a factory is similar to that illustrated for a warehouse in Table 7.4.

Table 7.5 A valuation of a licensed property based on the 'direct' method of valuation

Units	Volume figures	Price per unit (£)	£
680	680 draught barrels	5.35	3,638
240	120 bottled barrels	6.50	780
2,600	1,300 gals wine and spirits	0.90	1,170
3,520	Tied rent 3,520 units	0.90	3,168
	Total brewer's income		8,756
	Brewer's bid for rent (50% brewer's income)		4,378
	Catering receipts		
	Adjusted for 'tone' (£11,500 @ 10%)		1,150
	2 gaming machines		225
			5,753
	Rateable value		£5,750

The above is the valuation confirmed in *Jones (VO) v. Toby Restaurants South Ltd* (1992) which concerned appeals made in 1986 and 1989 against assessments entered into the 1973 valuation lists. The adjustment for 'tone' included in the valuation should not be necessary where quinquennial revaluations are maintained.

7.5.11   In addition, there are purpose-built industrial premises for which rental evidence is not available. Such properties tend to be owner-occupied and, in such a case, the valuation is achieved using the contractor's test (see 6.6).

7.5.12   Examples of industrial properties which may be valued using the contractor's test are steel works, chemical plants, oil refineries and cement works.

7.5.13   Within industrial hereditaments there is likely to be specialist plant and machinery and additional information is likely to be needed in order to identify the item of plant and machinery, establish if it is rateable (see 3.2.14–3.2.43) and ascribe to it an appropriate value.

7.5.14   Plant and machinery is valued using a replacement cost, to which the appropriate decapitalisation rate (currently 3.67% for non-educational or hospital hereditaments and 5.5% in all other cases) is applied.

7.5.15   The plant and machinery content of an average warehouse is about 2% of its rateable value and in a non-specialist factory plant and machinery accounts for about 5% of the assessment. For more specialised property types, plant and machinery may account for up to 30% of the rateable value and, in extreme cases, for over 90%. However, despite this range of impact on the rateable value, the presence of plant and machinery within buildings is almost invariable.

7.6   Licensed premises

7.6.1   Licensed premises, i.e. those with a licence to sell intoxicating liquor, are measured to net internal area and, traditionally, valued using a variant of the profits method, called the direct approach method, which follows the decision in *Robinson Brothers (Brewers) Ltd. v. Durham County Assessment Committee* (1938) that brewers could be considered as potential hypothetical tenants of public houses.

7.6.2   The direct approach method assumes that a public house can be occupied (in the rating context) by a brewery which would 'tie' the public house to sell specific products provided by that brewery. The brewery can, therefore, receive the profit both on the wholesale trade to the public house and on the retail trade from the public house to the consumers. Such relatively high levels of profit mean that a brewery could afford to outbid other hypothetical tenants in the market.

7.6.3   The direct method of valuation assumes, therefore, that the brewery is the hypothetical tenant (having installed a 'tied' tenant to run the public house). Firstly, it is necessary to establish the reasonably maintainable trade as at the valuation date.

7.6.4   It will depend on the nature of the trade within the premises whether there is 'wet' trade (i.e. the sale of liquor) and 'dry' trade (i.e. the sale of food) on the premises. The profitability of both kinds of trade is taken into account. Similarly, the profits from any gaming machines should also be included. An example of a valuation based on the 'direct' method of valuation is illustrated in Table 7.5.

7.6.5   There is, however, growing concern that the 'direct' method of valuation is not appropriate since the 1988 report of the Monopolies and Mergers Commission which required breweries to limit the number of tied public houses they hold. This has radically increased the market for public houses, which are being acquired by purchasers who are not brewers and for whom the 'direct' method of valuation is not suitable.

7.6.6 Methodologies, including a price per square metre overall, have been proposed but none has so far been widely adopted by the profession.

7.7 Petrol filling stations

7.7.1 Petrol filling stations are valued using a variant of the profits method called 'throughput'.

7.7.2 The 'throughput' variant of the profits method involves an estimate of future sales, based to some extent on current and past sales, and on the trade enjoyed by other stations in the locality.

7.7.3 Petrol filling stations are often included within out-of-town retail properties, particularly hypermarkets and motorway service stations. Many petrol filling stations include a small shop selling groceries and confectionery as well as goods for motor vehicles, and a car-wash; others, the more traditional kind, include showrooms, workshops and open yard space to store vehicles. These different ancillary uses are valued using comparable rental evidence.

7.7.4 The hypothetical tenant for a petrol filling station may be an independent trader or an oil company. If the petrol filling station sells petrol and petroleum products almost exclusively and is in a popular trading location, it is likely that the hypothetical tenant for the hereditament is an oil company, which will put a tied manager into the premises. In such a case, the rental bid which the hypothetical (oil company) tenant offers will be based on the tied rent receivable plus a proportion of the wholesale profit which the company would make from the sale of its products from the hereditament.

7.7.5 The valuation of a petrol filling station is illustrated in Table 7.6.

7.7.6 Factors which affect the value of petrol filling stations, over and above the accommodation, are: the nature of the road on which the station is sited; the extent of car ownership within the locality; the planning situation for the locality; the visibility of the station from the highway; ease of access; traffic volume and speed restrictions at the entrance to the station; the nature and range of products sold; opening hours and pricing policy; the nature and extent of personal

Table 7.6 A valuation of a petrol filling station

Estimated capital value		
Forecourt sales: 6,000,000 litres per annum @ 28.5p	£1,710,000	
Decapitalised: YP @ 9% in perpetuity	11.111	
Estimated rental value		£153,900
Shop at 50 sq. m. @ £130	£6,500	
Car wash: third of net profit	£12,000	
Total additional rental value		£18,500
Rental value		£172,400

Based on J. R. E. Sedgwick 'Garages and Petrol Filling Stations' in W. H. Rees (ed) *Valuation: Principles into Practice*, The Estates Gazette (4th ed., 1992) p. 450.

This is a valuation of a prime self-service petrol station on a busy trunk road in an outer west London area. Visibility is good and there is a 40 mph speed limit. The volume of traffic on the route averages 30,000 vehicles a day and there is a call-in rate at the station of ... 760. At an average purchase of [23 litres] this results in sales of [17,480 litres] per day or, ... say, [6,000,000 litres] a year on a six-and-a-half-day-week basis, ... this is ... the ascertained throughput for the previous year, sales having risen steadily over the last five years. The prices charged are normal for the area and credit cards are accepted. There is a forecourt shop and a car wash. The forecourt itself is covered by a canopy and the pumps are in 'starting line' formation to achieve maximum capacity.

service available; the physical nature and layout of the station, including physical and technological obsolescence of plant and machinery.

7.7.7   The valuation of petrol filling stations, whether for rating purposes or for any other purposes, is a specialism and a thorough knowledge of both the properties and the trades involved is vital for an accurate and professional valuation.

7.8   Bibliography

The following is a brief bibliography. Some other relevant texts are listed in the full Bibliography. Other specialist publications dealing with the general valuation principles applicable to specific property types, together with any changes in the specialist trades which affect the use and value of property should also be considered.

Marshall, Harvey and Williamson, Hazel, 1994. *Law and Valuation of Leisure Property*. The Estates Gazette Ltd.

Rees, W. H. (ed), 1992. *Valuation: Principles into Practice*. The Estates Gazette Ltd.

RICS/ISVA, 1993. *Code of Measuring Practice*. Surveyors Publications.

Chapter 8

Rating lists

8.1 Synopsis

8.1.1 Rating lists contain an entry for all hereditaments the occupiers or owners of which are liable to the Uniform Business Rate. The lists are produced every five years by valuation officers.

8.1.2 Rating lists are conclusive proof of liability to rates and rates are levied in accordance with the entries in the lists.

8.1.3 There are two kinds of rating lists: local non-domestic rating lists and two central non-domestic rating lists (one for England and one for Wales).

8.1.4 The local non-domestic rating lists contain entries for hereditaments which are located in the areas of billing authorities, one list for each billing authority.

8.1.5 The central non-domestic rating list contains entries for hereditaments which belong to such nation-wide hereditaments as electricity, gas, and railway companies.

8.2 Introduction

8.2.1 Section 41 (1) requires the valuation officer for each billing authority to compile and then maintain a local non-domestic rating list comprising all relevant non-domestic hereditaments (see Chapter 3) within that billing authority's area.

8.2.2 Similarly, s. 52 (1) requires the central valuation officer to compile and then maintain a central non-domestic rating list. (see 8.4 for further details of the central non-domestic rating list.)

8.2.3   The valuation officers are impartial officers appointed by the Commissioners of Inland Revenue employed by the Valuation Office Agency (VOA), with duties to compile and maintain rating lists for the areas for which they are responsible.

8.2.4   Lists are compiled every five years. The first rating lists were compiled on 1 April 1990 and the current lists (at the time of writing) on 1 April 1995. The rating lists come into force on the day on which they are compiled and remain in force until they are replaced by new lists.

8.2.5   *Daily liability*
              The concept of a daily rate liability was introduced by the Local Government Finance Act 1988. This is because the legislation was drafted in tandem with the Community Charge legislation, which imposes a daily liability to the Community Charge on community-charge payers (see Appendix A for details of the Community Charge).

8.2.6   Thus, each ratepayer is liable for the uniform business rate for each day on which the hereditament appears in the rating list.

8.2.7   Valuation officers make changes to the list and notify the occupier of that change (see Chapter 9).

8.3   Local non-domestic rating lists

8.3.1   For each billing (local) authority area, a valuation officer (VO) is appointed by the Commissioners of Inland Revenue.

8.3.2   Under s. 41 (1)1988 Act, the valuation officer has a statutory duty to compile and maintain a local non-domestic rating list for each billing authority for which he is appointed by the Commissioners of Inland Revenue.

8.3.3   The occupier of any relevant non-domestic hereditament shown in the rating list is liable to pay rates in respect of that hereditament for each day on which it appears in the rating list (s. 43 (1) 1988 Act). Similarly, the owner of any relevant hereditament shown in the rating list is liable to pay rates in respect of that hereditament for each day in which it appears in the rating list and for which no occupier pays rates (s. 45 (1) 1988 Act) (see also 2.4). Thus, the rating lists are conclusive proof of liability to rates.

8.3.4   Only hereditaments which are entirely exempt rates or which are entirely domestic are excluded from the list (see Chapter 4). Thus, the rating list contains relevant non-domestic hereditaments and composite hereditaments (see Chapter 3).

8.3.5   The current rating list became effective on 1 April 1995, when it was 'compiled', and is conclusive evidence on all matters concerning rate liability in respect of a property, other than the identity of the occupier.

8.3.6   The rating list is a daily list (see 8.2.5), and must show, for each day in each chargeable financial year for which it is in force, each hereditament which is:
(a)   situated in the billing authority's area;
(b)   a relevant non-domestic hereditament;
(c)   not entirely domestic property;
(d)   not entirely exempt from local non-domestic rating; and
(e)   not a hereditament which must be shown in a central non-domestic rating list (s. 42 (1) 1988 Act).

8.3.7   Any composite hereditament (i.e. a hereditament part of which comprises domestic property (s. 64 (9) 1988 Act)) and any hereditament part of which is exempt, must be identified as such in the list (s. 42 (2) 1988 Act).

8.3.8   Where a hereditament spans the border between two billing authorities, the hereditament is entered in the list for the billing authority area where the majority of the value is located (reg. 6 Non Domestic Rating (Miscellaneous Provisions) Regulations 1989 (SI 1989 No. 1060)). Guidance exists in the regulations for resolving any dispute.

8.3.9   Local non-domestic rating lists are required to show a description of each hereditament, its address, rateable value and any reference number ascribed to it by the valuation officer. In respect of any alteration directed to be made by a tribunal, the list must state whether the direction was given by the Valuation Tribunal (V) or the Lands Tribunal (L), and must also show the total of rateable values shown in the list in accordance with s. 42 (4) 1988 Act (ibid. regs. 2–4). Thus, the pages within the list have the headings shown in Figure 8.1:

Assessment Number	Description	Address	Rateable Value	Effective Date	V or L

Figure 8.1   Local non-domestic rating list

Note: V   indicates that an alteration was made as the result of
an order by a valuation tribunal
L   indicates that an alteration was made as the result of
an order by the Lands Tribunal

8.3.10   Since these matters must appear in the rating list, they are matters on which appeals can be made (see Chapter 9).

8.3.11   Rating lists are prepared every five years (s. 41 (2) 1988 Act). This requirement for quinquennial revaluations existed under the pre-1990 legislation, but was never in fact achieved in England and Wales. The first rating list took effect on 1 April 1990, the current list (at the time of writing) took effect on 1 April 1995 and the next rating list will take effect on 1 April 2000.

8.3.12   *Revaluations*
In the year preceding the coming into force of the list (e.g. 1994 for the 1995 list), the valuation officer compiles the local non-domestic rating lists in draft and, not later than 31 December of the year before the list takes effect (i.e. 31 December 1994 for the 1995 list), copies the draft list and hands the copy to the billing authority (s. 41 (5) 1988 Act).

8.3.13   The valuation officer retains and amends the draft list, informing the billing authority of the amendments by monthly schedules, until, on the following 1 April (1 April 1995 for the 1995 list), the list comes into force.

8.3.14   On receipt of the draft list, the billing authority is required to deposit it at its principal office and take such steps as it thinks suitable for giving notice of it (s. 41 (6) 1988 Act). When the valuation officer informs the billing authority that amendments have been made, the billing authority is required to alter the deposited copy accordingly (reg. 6 Non Domestic Rating (Miscellaneous Provisions) (No. 2) Regulations 1989 (SI 1989 No. 2303).

8.3.15   There is no requirement to inform the occupier at this stage (s. 55 (1) 1988 Act, and reg. 6 Non Domestic Rating (Miscellaneous

Provisions) (No 2) Regulations 1989 (SI 1989 No. 2303)), and ratepayers are not individually notified of their new assessments until they receive their rate demand notices in the financial year in which the list takes effect, e.g., the year beginning 1 April 1995 for the 1995 list.

8.3.16   The valuation date is two years before the date on which the list takes effect, i.e. 1 April 1993 (for the list which took effect on 1 April 1995), with the hereditament assumed to be in its state as at 1 April 1995 (for the 1995 list). All alterations are made as at the 1 April 1993 antecedent valuation date (for the 1995 list see 5.4.45).

8.3.17   Values may be altered by the valuation officer at any time after the date the list takes effect (1 April 1995 for the 1995 list), and billing authorities are informed of the alterations by monthly updates.

8.4   Central non-domestic rating lists

8.4.1   Under s. 52 of the 1988 Act, a central valuation officer, appointed by the Commissioners of Inland Revenue, compiles a central (as opposed to a local) non-domestic rating list. There is a central rating list for England and a central rating list for Wales. Central non-domestic rating lists were introduced '*with a view to secure the central rating en bloc of certain hereditaments*' (s. 53 (1) 1988 Act), e.g. railway hereditaments (see 8.4.10–8.4.17).

8.4.2   The Central Rating Lists Regulations 1989 (SI 1989 No. 2263) as amended, prescribe the contents of the central rating lists.

8.4.3   This list is produced centrally and the current central rating list (at the time of writing) came into force on 1 April 1995. As for local non-domestic rating lists, there is a statutory requirement for quinquennial revaluations (s. 52 (2) 1988 Act). The procedure to be followed is similar to that for local lists, except that the Secretaries of State for the Environment and for Wales are sent the copy of the central non-domestic rating list by the central valuation officer and they place it on deposit at their principal offices.

8.4.4   The central rating list contains the rateable values of hereditaments which span the country and, therefore, the areas of many billing authorities. Having one rateable value for the premises of each occupier allows for the rating of national networks and the former statutory undertakings (e.g. electricity and railway properties) to be achieved *en bloc*, and since all the rate revenue raised is paid into a central government 'pool', there is no need for the apportionment of rateable values between billing authorities.

8.4.5 The central rating list is compiled, maintained and held by the central valuation officer and the copy is held by the Secretary of State for the Environment at the Marsham Street offices of the Department of the Environment in London and by the Secretary of State for Wales at the Welsh Office in Cardiff.

8.4.6 **Content of the central rating list**
Section 53 (2) requires that the central rating list contains, for each day, the name of the designated person and the relevant hereditament which is occupied (or if unoccupied, owned) by him and the rateable value.

8.4.7 The central rating list may also contain such information about the hereditaments as may be prescribed by the Secretary of State in regulations.

8.4.8 The 1989 regulations (SI 1989 No. 2263, as amended) designate the person liable to pay the rates (under s. 53 (1)); prescribe in relation to him a description of relevant non-domestic hereditaments occupied or, if unoccupied, owned by him; give the registered office of the designated person and, if a registered company, its registration number; and give the first day (if later than 1 April 1995) for which the rateable value shown against the designated person took effect.

8.4.9 'Relevant hereditaments' are valued in accordance with the appropriate regulations prescribed by the Secretary of State (The Central Rating Lists Regulations 1994 (SI 1994 No. 3121)), (see also 6.3), with 'excepted hereditaments' being valued in the normal way.

8.4.10 **Canal hereditaments**
The designated person is the British Waterways Board (ibid. part 1 to the Schedule). Relevant hereditaments are defined to include waterways, reservoirs, lighthouses, docks, waste disposal tips; and excepted hereditaments include premises so let out as to be capable of separate assessment, a shop, warehouse and office premises (Non Domestic Rating (Railways, Telecommunications & Canal) Regulations 1994 (SI 1994 No. 3123)).

8.4.11   *Electricity supply hereditaments (see 6.3.5)*
The designated persons include National Power plc, the Power Generation Company plc, the National Grid Company plc, South Wales Electricity plc, Manweb plc (part 2 to the Schedule in the Central Rating Lists Regulations 1994 (SI 1994 No. 3121)). Relevant hereditaments include those used wholly or mainly for the generation, transformation and transmission of power, or for the purposes of the functions of a public electricity supplier. Note that hereditaments occupied for the purposes of generating electricity by means of tidal flow are not included on the central rating list. Excepted hereditaments include a shop, showroom and office premises not on occupational land (see also Electricity Supply Industry (Rateable Values) Order 1994 (SI 1994 No. 3282), as amended).

8.4.12   *Gas hereditaments (see 6.3.8)*
The designated person is British Gas plc (part 3 to the Schedule in the Central Rating Lists Regulations 1994 (SI 1994 No. 3121)). Relevant hereditaments include those used wholly or mainly for the purposes of a public gas supplier, and for the supply, installation or maintenance of gas appliances. Excepted hereditaments include premises used wholly or mainly for the manufacture of plant and gas fittings, a shop, showroom and office premises not situated on occupational land (see also British Gas plc (Rateable Values) Order 1994 (SI 1994 No. 3283)).

8.4.13   *Railway hereditaments (see 6.3.10)*
The designated persons include the British Railways Board, Railtrack plc and London Underground Ltd. (part 4 to the Schedule in the Central Rating Lists Regulations 1994 (SI 1994 No. 3121)). Relevant hereditaments include those used wholly or mainly for the carriage or goods or persons by rail, and ancillary purposes, including the purpose of exhibiting advertisements. Excepted hereditaments include a shop, hotel, place of public refreshment, offices not situated on operational land and premises so let out as to be capable of separate assessment (see The Non Domestic Rating (Railways, Telecommunications & Canal) Regulations 1994 (SI 1994 No. 3123) and the Railways (Rateable Values) Order 1994 (SI 1994 No. 3284).

8.4.14   *Telecommunications hereditaments*
The designated persons are British Telecommunications plc and Mercury Communications Ltd. (part 5 to the Schedule in the Central Rating Lists Regulations 1994 (SI 1994 No. 3121)). Relevant hereditaments are all hereditaments occupied by posts, wires, underground cables and ducts, telephone kiosks, switchgear

and other equipment not within a building, or by easements or wayleaves, which is used for telecommunication services (see also reg. 4 of the Non Domestic Rating (Railways, Telecommunications and Canal) Regulations 1994 (SI 1994 No. 3123)). Note that such hereditaments are no longer valued using a statutory formula (see 6.3.3).

8.4.15 *Water supply hereditaments (see 6.3.11)*
Designated persons include Dwr Cymru Cyfyngedig, Severn Trent Water Ltd., Anglian Water Services Ltd. (part 6 to the Schedule in the Central Rating Lists Regulations 1994 (SI 1994 No. 3121)). Relevant hereditaments are used wholly or mainly for the purposes of a water undertaker. Excepted hereditaments include premises used wholly or mainly for the manufacture, storage, sale, display or demonstration of apparatus or accessories for consumers, and office premises not on operational land (see also Water Undertakers (Rateable Values) Order 1994 (SI 1994 No. 3285)).

8.4.16 *Long-distance pipelines*
Designated persons include Mainline Pipelines Ltd., BP Chemicals Ltd., British Steel plc, Shell Chemicals UK Ltd. (part 6 to the Schedule in the Central Rating Lists Regulations 1994 (SI 1994 No. 3121)). Relevant hereditaments include cross-country pipelines situated within the area of more than one billing authority.

8.4.17 The rateable values of each of the kinds of rateable hereditaments described above (8.4.10 to 8.4.16) are assessed in accordance with separate statutory instruments (see 6.3).

8.5 Inspection

8.5.1 There is a right to inspect the current local rating list held by the valuation officer (Sch. 9, para. 8 (1) 1988 Act) or the copy of the list and proposed lists deposited with the billing authorities (ibid. para. 8 (2) and (4)).

8.5.2 Also, it is possible to transcribe information from the lists at no charge and to obtain photocopies at a reasonable charge (ibid. para. 8 (9)).

8.5.3 There is a similar right to inspect, etc., the copy of the central non-domestic rating list (ibid. para. 8 (3) and (5)).

8.6   Information

8.6.1   Billing authorities and precepting authorities are under a statutory duty to inform the valuation officer about material changes in circumstances. The valuation officer may either alter the list within a specified time or inform the authority in cases where he considers that an alteration is not justified (see 9.3.4–6).

8.7   Rent returns

8.7.1   In preparing a new rating list, the valuation officer is empowered to request information and to enter and inspect premises (Sch. 9 para. 7 1988 Act). The valuation officer may serve a notice (issued under Sch. 9 para. 5) on the owner, the occupier or both requesting information which the valuation officer 'reasonably believes' (ibid.) will assist in the carrying out of his functions.

8.7.2   Such written details include information regarding the rent paid on any hereditament, which may be used as evidence on which to base a rateable value. See also 10.13 for the use of such information in valuation proceedings.

8.7.3   If a person served with a notice fails, without reasonable excuse, to comply within 21 days, or if, in supplying information in purported compliance, he makes a statement which he knows to be false, he will be liable to a fine, or to both a fine and a term of imprisonment (Sch. 9, para. 5).

8.8   Check-list

8.8.1   There are two kinds of rating lists: the local non-domestic rating list (one prepared for each billing authority area) and the central non-domestic rating list (containing hereditaments valued *en bloc*, e.g. railway hereditaments) (8.3 and 8.4).

8.8.2   The valuation officer is responsible for the preparation, maintenance and alteration of both the local and the central rating lists (8.2.1–8.2.3)

8.8.3   Lists are compiled every five years and the current list (at the time of writing) came into force on 1 April 1995 (8.2.4).

8.8.4   The lists must be in the form prescribed and include all hereditaments which are neither totally domestic nor totally exempt (8.3.4).

8.8.5   Returns can be required from occupiers and owners for information which the VO reasonably believes will assist in the carrying out of his duties (8.7).

Chapter 9

Appeals

9.1 Synopsis

9.1.1 Alterations to the non-domestic rating lists are made by valuation officers as and when necessary to maintain the accuracy of the lists.

9.1.2 Other 'interested persons' may make proposals to alter the lists in a standard form following a set procedure.

9.1.3 Proposals can be 'well-founded' and the list altered as a result; they can be withdrawn; parties can reach an agreement on a new entry; or the matter can be determined at a hearing by the valuation tribunal.

9.2 Rights of appeals

9.2.1 This Chapter considers the rights and procedures involved in appealing against an entry in the rating lists. In addition to such an appeal against the entries in the rating lists, it is possible to appeal against:
(a)   the UBR multiplier, for judicial review (under s. 138, 1988 Act);
(b)   the rates paid on the grounds of hardship (under s. 49, 1988 Act), at the discretion of the billing authority; and
(c)   a liability order, etc. (para. 3 Sch. 9 1988 Act) to the magistrates' court (see Chapter 11).

These appeals are not considered further.

9.2.2 The rating lists are conclusive on matters of value, effective date of a change of value, completion date and entitlement to an exemption (under Sch. 5). For this reason, these matters cannot be challenged in rate-recovery procedures and must, therefore, be challenged as an appeal against the list itself.

9.3   Proposals to alter the local non-domestic rating lists

9.3.1   The way in which entries in the local non-domestic rating list are altered is by a proposal to alter the rating list.

9.3.2   The procedure dealing with such proposals to alter the local rating lists is contained in the Non Domestic Rating (Alternation of Lists and Appeals) Regulations (SI 1993 No. 291) (the 1993 regulations), as amended by the Non Domestic Rating (Alteration of Lists and Appeals) (Amendment) Regulations (SI 1995 No. 609) (the 1995 regulations), made under s. 55 of the 1988 Act. This procedure applies to proposals to alter the rating list made on or after 1 April 1995.

9.3.3   Those who make proposals, billing authorities (see 9.3.7–8) and interested persons (see 9.3.9 *et seq.*), need to have 'reason to believe' that at least one of the grounds on which it is open to them to make a proposal exists (reg. 4A (2) 1995 regulations).

9.3.4   *Alteration of the list by valuation officers*
Section 41 (1) 1988 Act gives the valuation officer the specific duty to compile and maintain the list. There is, however, no explicit power to alter the list – this must obviously be inferred, since reg. 18 of the 1993 regulations states that, having made the alteration, the valuation officer must serve notice of an alteration within a specified time limit. There is, however, no sanction against him should he fail to do so.

9.3.5   The valuation officer, therefore, has no statutory power to make a proposal to alter the rating list. However, there is no need for the valuation officer to make a proposal to alter the rating list. Since the valuation officer holds the rating list, if he has reason to believe that entries in the list are incorrect, he can merely make necessary alterations and inform those interested (relevant authorities and the occupier or a rateable owner) that he has done so. If they disagree with this alteration, they can make a proposal to alter the rating list themselves. Ratepayers are not informed if the alteration is merely to correct a clerical error but billing authorities must be informed of all alterations within four weeks, so that they can alter their copy of the list as soon as is reasonably practicable (reg. 18 of the 1993 regulations, as amended).

9.3.6 Once the entry in the list is amended (this must include the effective date of the amendment) rates must be paid on the altered rateable value as from the effective date.

9.3.7 *Alteration of the list by billing authorities*
The billing authority has no power to make a proposal on a property on which it is not a ratepayer, except in the circumstances outlined in 9.3.8.

9.3.8 Billing authorities may make a proposal for the alteration of the list at any time during the life of the list for the following reasons:
(a) the rateable value shown in the list for the hereditament is inaccurate by reason of a material change of circumstances which occurred on or after the day on which the list was compiled;
(b) the rateable value or any other information shown in the list for the hereditament is shown, by reason of a decision in relation to another hereditament of a valuation tribunal, the Lands Tribunal or a court determining an appeal or application for review from either such tribunal, to be or to have been inaccurate;
(c) a hereditament not shown in the list ought to be shown in that list;
(d) a hereditament shown in the list ought not to be shown in that list;
(e) the list should show that some part of a hereditament which is shown in the list is domestic property or is exempt from non-domestic rating, but does not do so;
(f) the list should not show that some part of a hereditament which is shown in the list is domestic property or is exempt from non-domestic rating, but does so.
(Reg. 4A of the Non Domestic Rating (Alteration of Lists and Appeals) (Amendment) Regulations (SI 1995 No. 609) which amends reg. 4 Non Domestic Rating (Alternation of Lists and Appeals) Regulations (SI 1993 No. 291).)

9.3.9 *An 'interested person'*
An 'interested person' is defined (reg. 2 (1) of the 1993 regulations) as an occupier; a person (other than a mortgagee not in possession) having either a legal or an equitable interest in any part of the hereditament that would entitle him to possession (after the cessation of any prior interest); and any person having a 'qualifying connection' with the occupier or other 'interested person'. 'Qualifying connection' reflects the legal connection

which may exist between, for example a parent company and its subsidiary, or between subsidiary companies.

9.3.10   An 'interested person' can request to be made a party to proceedings arising from a proposal which was not made by him but which relates to his hereditament, provided that he serves notice on the valuation officer within two months of the service of the proposal (reg. 11 (e) (ii) 1993 regulations as amended by the 1995 regulations).

9.3.11   An 'interested person' can make a proposal under the following circumstances:

(a)   the rateable value shown in the list for a hereditament was inaccurate on the day the list was compiled;

(b)   the rateable value shown in the list for the hereditament is inaccurate by reason of a material change of circumstances which occurred on or after the day on which the list was compiled ('material change of circumstances' is defined in 9.10).

(c)   the rateable value shown in the list for the hereditament is or has been inaccurate by reason of an alteration made by a valuation officer;

(d)   the rateable value or any other information shown in the list for the hereditament is shown, by reason of a decision in relation to another hereditament of a valuation tribunal, the Lands Tribunal or a court determining an appeal or application for review from either such tribunal, to be or to have been inaccurate;

(e)   the day from which an alteration is shown in the list as having effect is wrong;

(f)   a hereditament not shown in the list ought to be shown in that list;

(g)   a hereditament shown in the list ought not to be shown in that list;

(h)   the list should show that some part of a hereditament which is shown in the list is domestic property or is exempt from non-domestic rating, but does not do so;

(i)   the list should not show that some part of a hereditament which is shown in the list is domestic property or is exempt from non-domestic rating, but does so;

(j)   property which is shown in the list as more than one hereditament ought to be shown as one or more different hereditaments;

(k)   property which is shown in the list as one hereditament ought to be shown as more than one hereditament;

(l)   the address shown in the list for the hereditament is wrong;

(m)   the description shown in the list for a hereditament is wrong; or

(n) any statement required to be made about the hereditament (under s. 42) has been omitted from the list.
(Reg. 4A of the 1995 regulations.)

9.3.12 Such proposals can be made at any time during the life of the list, strictly, at any time before the first anniversary of the compilation of the next list (reg. 4B(1) inserted by the 1995 regulations).

9.4 Form and content of proposal

9.4.1 A valid proposal must be in writing and served on the valuation officer. It must state the name and address of the person making it, including the capacity in which that person is acting, e.g. occupier or owner. It must identify the property and proposed manner in which the list should be altered. The proposal should contain a statement of reasons explaining why it is considered that the list is incorrect or, if there has been a material change of circumstances, the nature and date of the change, etc. (reg. 5A of the 1995 regulations).

9.4.2 In addition, the proposal must be signed (a typed name is insufficient), although it is possible for an agent to sign on behalf of an 'interested person' principal.

9.4.3 The proposal may relate to more than one hereditament provided that (ibid. reg. 5A (3) (b)):
(a) it is made on grounds which relate to whether or not the property comprises one or several hereditaments ((j) or (k) above – see 9.3.11); or
(b) where the maker of the proposal does so in the same capacity in respect of each hereditament; or
(c) where each hereditament is within the same building or curtilage.

9.4.4 The proposal establishes the matter(s) which are to be discussed between the maker of the proposal and the valuation officer, which could develop into the basis for a court case. However, in *Thomas (S) & Co. (Nottingham) Ltd v. Emett (VO)* (1955), *Grover's Trustees v. Corser (VO)* (1963) (cf *Daniels v. Peak (VO)* (1964)), the Lands Tribunal held that where a proposal was to increase an assessment on the grounds of structural alterations, the tribunal could review the assessment for the entire hereditament and was not limited to considering the value of the additions alone. This is reasonable in that the decision of the court is a rateable value for the hereditament which should be correct at the time the decision is made.

9.4.5   The facts established in *R. v. Winchester Area Assessment Committee* (1948) still apply. Thus the proposal must give sufficient information to enable the court to know:
(a)   whether an increase or a decrease is asked for;
(b)   to which of the existing entry or entries the proposal refers; and
(c)   what is the ground of complaint against that entry.

9.4.6   The judge in that case stated (at p. 555) that 'it is enough to state "incorrect or unfair" [although unfairness is no longer valid grounds for a proposal], unless there is some unusual ground, in which case it ought to be specified.'

9.4.7   However, it is of vital importance that the grounds of the proposal are sufficient to cover the points on which the case is to be based and yet specific enough not to result in the proposal being declared invalid.

9.5   Validity

9.5.1   If the valuation officer considers that a proposal has not been validly made, then he has four weeks after it was served on him to serve an invalidity notice on the person making the proposal stating why it is considered to be invalid. He must also draw attention to recipient's right to make another proposal (except in certain specified circumstances: see 9.5.2 and 9.5.4) or to appeal within four weeks to the valuation tribunal. (Reg. 7 of the 1993 regulations 1993 as amended by the 1995 regulations.)

9.5.2   If, on the second making, the proposal is again invalid, there is no opportunity to make another proposal (ibid.).

9.5.3   An appeal against the invalidity notice is initiated by service of a notice of disagreement on the valuation officer within four weeks (unless the valuation officer withdraws the invalidity notice within that time). The appeal takes the form of a transfer of all relevant documentation by the valuation officer to the clerk of the relevant valuation tribunal and must be determined by the valuation tribunal before the tribunal considers the appeal against the valuation officer's (deemed or actual) refusal to alter the list in response to the proposal.

9.5.4   The procedure for serving a fresh proposal for the alteration of a local rating list where the valuation officer has challenged the validity of a proposal is not available where the original proposal did not comply with the time limits (reg. 7 (4) Non Domestic Rating (Alteration of Lists and Appeals) Regulations 1993 (SI 1993 No. 291).

9.6   Proposal procedure

9.6.1   Four weeks after the proposal has been served on him, the valuation officer must acknowledge receipt of the proposal and serve copies of it on the ratepayer and, in certain circumstances, the billing authority (reg. 8, 1993 regulations, as amended).

9.6.2   There are now four possible outcomes to the procedure, as shown in figure 9.1:
(a)   the grounds of proposal will be agreed by the valuation officer, who will accept the proposal as 'well-founded' and alter the list accordingly (9.6.4);

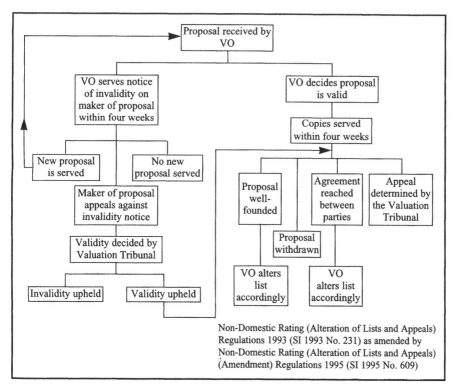

Non-Domestic Rating (Alteration of Lists and Appeals) Regulations 1993 (SI 1993 No. 231) as amended by Non-Domestic Rating (Alteration of Lists and Appeals) (Amendment) Regulations 1995 (SI 1995 No. 609)

Figure 9.1   Procedure for proposals made to alter the local rating lists

(b)   the maker of the proposal will withdraw the proposal and the appeal procedure ends (9.6.5);

(c)   the valuation officer will object to the list being altered in accordance with the proposal, the parties will come to an agreement as a result of negotiation and the list will be altered accordingly (9.6.8); or

(d)   the valuation officer will object to the list being altered in accordance with the proposal, the parties will fail to reach an agreement as a result of negotiation and the case will be decided by the valuation tribunal (see 9.7 and Chapter 10).

9.6.3   These possible solutions are described below in more detail.

9.6.4   *Proposal well founded*

If the valuation officer agrees with the proposal, he serves a notice accordingly on the person making the proposal and alters the list as soon as is reasonably practicable (reg. 9 of the 1993 regulations, as amended by reg. 7 of the 1995 regulations).

9.6.5   *Withdrawal of proposal*

A proposal may be withdrawn by the person who made it by notifying the valuation officer in writing of the withdrawal (reg. 10, ibid.).

9.6.6   Where the maker of the proposal is no longer the occupier, he must obtain the written agreement of the current occupier (ibid.). Thus, a proposal can be withdrawn only by the person who made it, with the agreement of any subsequent occupier.

9.6.7   If an 'interested person' has notified the valuation officer that he wants to be a party to the proceedings, the valuation officer is required to serve a notice of withdrawal on that 'interested person', who then has six weeks to serve notice on the valuation officer to the effect that he wishes to take over the proceedings in the place of the person who made the proposal (reg. 10 ibid.).

9.6.8   *Agreement*

Where a proposal is not withdrawn nor accepted as well founded, it often happens that the parties involved in the process agree to a different entry to that originally proposed. When all such parties sign the appropriate agreement forms, the valuation officer alters the list within two weeks of the date of the agreement (reg. 9 Non Domestic Rating (Alteration of Lists and Appeals) (Amendment) Regulations 1995 (SI 1995 No. 609), amending reg. 11 (1) (a) of the Non Domestic Rating (Alteration of Lists and Appeals) Regulations 1993 (SI 1995 No. 291)).

9.6.9   Agreement forms require the signatures of the maker of the proposal; the occupier at the date of the proposal; the occupier at the date of settlement; any other 'interested person' who could have made the proposal (e.g. a rateable owner); the billing authority, where, at the date of the proposal it could have made the proposal; and the valuation officer.

9.7   Appeal to the valuation tribunal

9.7.1   Where the proposal is not withdrawn, nor accepted as well founded, nor withdrawn, nor is agreement reached, an appeal is made by the maker of the proposal against the refusal of the valuation officer to alter the list.

9.7.2   The appeal takes the form of a transmission by the valuation officer of all relevant documentation to the clerk to the relevant valuation tribunal, and this occurs within three months of the date the proposal was served on the valuation officer (reg. 12 of the 1993 regulations, amended by reg. 10 of the 1995 regulations).

9.7.3   It continues to be possible for the parties involved to reach one of the other solutions (i.e. proposal is well founded, proposal is withdrawn, agreement is reached to an amended entry)' mentioned above. However, in their absences the transmission of the papers to the clerk to the relevant valuation tribunal will ultimately result in the case being listed for hearing by the valuation tribunal and the matter will be determined by the court (see Chapter 10).

9.8   Effective date

9.8.1   The rating list must show the date from which the alteration to an entry takes effect. This effective date is different, depending on the reason for altering the list.

9.8.2   The effective date for the following is the date on which the circumstances giving rise to the alteration occurred:
(a)   inclusion or deletion of a hereditament from the list, including the merging of several hereditaments or the splitting of a single hereditament;
(b)   a hereditament includes domestic accommodation or ceases to include domestic accommodation;

(c)   a hereditament becomes exempt, or is required to be shown in a central rating list, or ceases to be part of an authority's area;
(d)   a material change of circumstances.
(Regs. 15 and 16 of the 1993 regulations.)

9.8.3   Alterations made to correct inaccuracies in a list have effect on various days, depending on the circumstances.

9.9   Proposals to alter the central non-domestic rating list

9.9.1   Central rating lists may be altered by the central valuation officer at any time in order to fulfill his statutory obligation to compile and maintain a central rating list. The central valuation officer is also required to alter the central rating list in accordance with regulations made under s. 53 1988 Act.

9.9.2   'Designated persons' may make a proposal for the alteration of the list at any time. The 'designated person' is defined (reg. 20 Non Domestic Rating (Alteration of Lists and Appeals) Regulations 1993 (SI 1993 No. 291) as the person designated by regulations under s. 53 of the 1988 Act in relation to the description of hereditaments to which the entry relates. See 8.4.10–8.4.16 for 'designated persons'.

9.9.3   Designated persons may make a proposal to alter the list at any time if they consider that, in respect of any of their designated hereditaments, the rateable value shown in the list is not the amount properly determined, in accordance with the appropriate order (reg. 24 (1) of the 1993 regulations). A copy of the proposal is sent to the appropriate Secretary of State (Environment or Wales).

9.9.4   There is an opportunity (ibid.) for such proposals to be dealt with in the same way as proposals to alter the local rating lists (see 9.6) i.e.:
(a)   the grounds of proposal will be agreed by the valuation officer who will accept the proposal as 'well founded' and alter the list accordingly;
(b)   the maker of the proposal will withdraw the proposal and the procedure ends;
(c)   the valuation officer will object to the list being altered in accordance with the proposal, the parties will come to an agreement as a result of negotiation and the list will be altered accordingly; or
(d)   the valuation officer will object to the list being altered in

accordance with the proposal, the parties will fail to reach an agreement as a result of negotiation and the case will be decided by the valuation tribunal (see Chapter 10).

9.9.5   Those hereditaments required to be shown in the central rating list, the rateable values of which are prescribed, will not have their entries altered in this way.

9.9.6   Proposals to alter the central rating list must be made in writing, stating the designated person's name and address, show the proposed rateable value, give the reasons for believing that the list is incorrect and state any proposed new effective date. The proposal must be served on the central valuation officer (reg. 24 (1) of the 1993 regulations).

9.9.7   Within four weeks of altering a central list, the central valuation officer must notify the designated person and the appropriate Secretary of State (for the Environment or for Wales). The Secretary of State will alter the copy deposited at his principal office (s. 52 (6B) of the 1988 Act and reg. 23 (1) of the 1993 regulations as amended).

9.10   Material changes in circumstances

9.10.1   Schedule 6, para. 2 (7), lists a number of circumstances which can cause changes justifying a proposal to alter the rating list.

9.10.2   These circumstances are :
(a)   matters affecting the physical state or physical enjoyment of the hereditament;
(b)   the mode or category of occupation of the hereditament;
(c)   the quantity of refuse or waste material which is brought onto and permanently deposited on the hereditament;
(d)   the quantity of minerals or other substances in or extracted from the hereditament;
(e)   matters affecting the physical state of the locality in which the hereditament is situated or which, though not affecting the physical state of the locality, are nonetheless physically manifest there; and
(f)   the use or occupation of other premises situated in the locality of the hereditament.

9.10.3   This wording stems from s. 121 of the 1988 Act, which was designed to amend the effect of s. 20 of the General Rate Act 1967 so as to reverse the decision of the House of Lords in *Clement (VO) v. Addis Ltd* (1988).

9.10.4   It was held in *Addis* (ibid.), that, to effect a change in the rateable value, a material change did not necessarily have to be a physical change, so that intangibles, such as an enterprise zone created as a result of legislation were a 'material change' which could be taken into account in rating valuation. In response to that decision, the government sought to change the law so that the only changes that could be taken into account were changes of a physical kind. This they have succeeded in doing, in the above section, and the only material changes in circumstances which can now be reflected in the alteration of a rating assessment are those listed in 9.10.2 above.

9.11   Certificates of value

9.11.1   The valuation officer has power to issue certificates of value, which are required for the purposes of calculating transitional relief (Sch. 7A, paras. 10–12 1988 Act). The values certified are values under the previous list which are used as a basis for calculating any transitional relief (see 4.6.4–16).

9.11.2   An 'interested person' may give notice to the valuation officer of his dissatisfaction with a certificate within six months of receipt (reg. 30 (1) Non Domestic Rating (Alternation of Lists and Appeals) Regulations 1993 (SI 1993 No. 291)).

9.11.3   If, within four weeks of the date of the notice to the valuation officer, the appeal has not been withdrawn and the parties have not reached an agreement, the 'interested person' is deemed to appeal to the valuation tribunal.

9.11.4   The appeal is instituted by serving notice on the valuation officer giving reasons for dissatisfaction, and unless the matter is agreed within four weeks, the disagreement is referred to the valuation tribunal (see 10.15).

9.12   Completion notices

9.12.1   A completion notice is served by a billing authority if the authority believes that any work necessary to complete a building under construction can be finished within three months (see

2.4.28–2.4.46). At the end of that period of time, the building is deemed to be rateable and rates levied accordingly (see 2.4.28–2.4.41).

9.12.2 Appeals against completion notices must be made in writing (notice of appeal) within four weeks of their service and must be accompanied by a copy of the completion notice and a statement of the grounds on which it is made (reg. 29, of the 1993 regulations).

9.12.3 Such appeals are heard by the valuation tribunal (see 10.14).

9.13   Check-list

9.13.1 Appeals can be made against entries in the rating list by the billing authority and 'interested persons', who are able to make a proposal to alter the rating list during the life of the list (9.3).

9.13.2 The proposal must be in a specified form in order to be valid (9.4 and 9.5).

9.13.3 Once a proposal has been made, it must be disposed of by one of the following:
(a)   being treated as 'well founded' (9.6.4);
(b)   being withdrawn (9.6.5–9.6.7);
(c)   by agreement (9.6.8–9.6.9); or
(d)   by an appeal to the valuation tribunal (9.7).

9.13.4 The rating list must show the effective date, i.e. the date on which an alteration takes effect (9.8).

9.13.5 Proposals to alter the central rating list follow a similar procedure (9.9).

9.13.6 Proposals can be made to alter the list because of a material change in circumstances (9.10).

9.13.7 The valuation officer can issue certificates of value which are required for the implementation of transitional arrangements (9.11).

9.13.8 Appeals against completion notices are made to the valuation tribunal (9.12).

Chapter 10

Valuation tribunals

10.1 Synopsis

10.1.1 Valuation tribunals are the court of the first instance in resolving a dispute for rating purposes.

10.1.2 Appeals can be dealt with by written representation or a hearing (possibly preceded by a pre-hearing review).

10.1.3 Appeals from the decisions of valuation tribunals go to the Lands Tribunal and thence, on a point of law only, to the Court of Appeal.

10.2 Origins and jurisdiction

10.2.1 Valuation tribunals were renamed and their powers extended by s. 15 of the Local Government Finance Act 1992 (formerly, they were called Valuation and Community Charge Tribunals, with all the responsibilities and jurisdiction imposed on those courts by the Valuation and Community Charge Tribunals (Transfer of Jurisdiction) Regulations 1989 (SI 1989 No. 440)).

10.2.2 Thus, under the 1989 Transfer of Jurisdiction Regulations (ibid.), with effect from 1 May 1989, valuation tribunals have jurisdiction to hear and determine appeals and applications in connection with pre-1 April 1990 rating matters, and under the Valuation and Community Charge Tribunal Regulations 1989 (SI 1989 No. 439), they have jurisdiction to deal with appeals resulting from the 1990–93 system of community charges (poll tax) (see Appendix A), and certain appeals against the Council Tax (see Chapter 14).

10.2.3   For the purposes of the Uniform Business Rate, they determine appeals to both local and central rating lists, completion dates on completion notices and valuation officer's certificates in transitional phasing.

10.2.4   Their composition, jurisdiction, etc. are established under Sch. 11 of the 1988 Act (as amended by the Local Government Finance Act 1992 s. 15 and by Part VI of the Non Domestic Rating (Alteration of Lists and Appeals) Regulations 1993 (SI 1993 No. 291)).

10.3   Appeals to the valuation tribunal

10.3.1   The following are the only categories of appeal permitted under Schedule 11 of the 1988 Act:

(a)   appeals against valid proposals, i.e., to determine the correct rateable value and the effective date of the amendment, as specified in regulations made under s. 55, 1988 Act. Alterations to local lists are made under part II, and to the central list under reg. 28;

(b)   disputes regarding the validity of proposals (see 9.5);

(c)   appeals against valuation officer certificates relating to the phasing provisions which concern the regulations regarding certification of a hereditament's value by the valuation officer, under reg. 30 1993 regulations (see also 9.11). The appeal must be made within six months of the date of the certificate and may be considered by the Lands Tribunal on appeal from the valuation tribunal. The valuation tribunal can require the valuation officer to review his decision;

(d)   appeals against completion notices (Sch. 4A 1988 Act (see also 9.12)). On receipt of a completion notice, the VO makes an alteration in the list and gives notice of this alteration. An appeal to the valuation tribunal may ensue under reg. 29 1993 regulations;

(e)   appeals against certain Community Charge matters, specified in s. 23 (2) 1988 Act, and against certain Council Tax matters, specified in s. 16 and s. 24 and Sch. 3 of the 1992 Act;

(f)   appeals against the central rating list are dealt with by the valuation tribunal in area in which valuation officer is located. Once an appeal has been withdrawn, the jurisdiction of the valuation tribunal ceases.

10.3.2 The valuation tribunal has the power to determine the correct rateable value for a hereditament, i.e., if a proposal is made to decrease an already under-assessed rateable value, the tribunal has the power to increase the assessment.

10.3.3 As explained previously (9.4.4), the valuation tribunal has the power to determine the correct rateable value for the whole hereditament when considering a proposal to increase an assessment following structural alterations. Reg 44 (4) Non Domestic Rating (alterations of the list and Appeals) Regulations 1993 requires that where the valuation tribunal determines a rateable value greater than that in the list or contended for in the proposal, the increased assessment has effect from the date of the court's decision.

10.3.4 The valuation tribunal also has the power to determine that an altered rateable value takes effect for a specified period of time (reg. 44 (6) Non Domestic Rating (Alterations of the List and Appeals) Regulations 1993 (SI 1993 No. 291)).

10.4 Composition of the tribunal

10.4.1 When a valuation tribunal convenes to decide cases, it comprises (normally) three members from a group of tribunal members appointed by the Secretary of State (for the Environment in England and Wales in Wales), for that particular locality (Sch. 11 1988 Act). The members choose their own president and each tribunal has a professional clerk experienced in rating matters.

10.4.2 Tribunal members are not required to be professionally qualified in rating valuation and law, and they tend to be lay members of the community, who are unpaid for their services, except for their expenses.

10.4.3 The tribunal consists of three members chosen from a panel. One of the members will be chairman or one of the deputy chairmen of the panel. The decision reached is that of the majority.

10.4.4 With the agreement of all parties present, the appeal may proceed with only two members, but if they cannot agree, the case must be re-listed for a future hearing.

10.4.5 It is the responsibility of the president of the tribunal to make suitable arrangements to ensure that appeals are heard in accordance with the relevant legislation.

10.5   Written representations and pre-hearing review

10.5.1   Appeals may be dealt with by written representations if all parties agree (reg. 35 (1) Non Domestic Rating (Alteration of Lists and Appeals) Regulations 1993 (SI 1993 No. 291). The procedure takes twelve weeks.

10.5.2   Written representations require the clerk to the valuation tribunal to notify all parties accordingly and, within four weeks, each party serves on the clerk written notice either of the reasons (or further reasons) supporting his case, or that he does not intend to make further representation.

10.5.3   Any such notice received is copied and sent to the other parties to the appeal together with a statement of the procedure. Each party must respond within four weeks to the notice served by the other parties. The responses are sent to the clerk, who copies them and sends them to the other parties.

10.5.4   Within four weeks of this exchange of written cases, the clerk presents all written statements to the valuation tribunal which is constituted to hear the appeal.

10.5.5   The valuation tribunal has the power to require parties to specify the grounds relied on, relevant facts and contentions. It can order that the appeal be disposed of by a normal hearing.

10.5.6   The valuation tribunal also has the power to require a pre-hearing review (ibid. reg. 36) in order to 'clarify the issues to be dealt with at a hearing'. Four weeks' notice must be given to the parties in advance of a review.

10.6   Parties

10.6.1   Parties to the appeal are the maker of the proposal (the appellant), the valuation officer and any of the following who challenge the validity of the proposal: the occupier at the date of the proposal; the occupier at the date of the hearing; the owner and others with a superior interest; and the billing authority.

10.6.2   There is no compulsion to attend, and the tribunal can hear an appeal in the absence of any party. Parties can appear in person or be represented by counsel or any other representative.

10.6.3   However, if a party to the appeal (other than the valuation officer) does not attend the hearing, the tribunal may dismiss the appeal, or they may decide to hear the case in the absence of the party (ibid. reg. 40 (4)). In addition, no right of appeal from the decision of the valuation tribunal exists for a party which fails to attend the valuation tribunal hearing (ibid. reg. (47 (2))).

10.6.4   The hearing is normally open to the public, unless, on application from a party, the valuation tribunal believes that his interests will be prejudicially affected by a public hearing (ibid. 40 (3)). Such an application may be accepted when evidence includes the presentation of accounts.

10.7   Before the hearing

10.7.1   The clerk to the valuation tribunal must give a minimum of 28 days' notice of the place and date of the hearing to all parties, although six weeks' notice is not unusual in practice.

10.7.2   In addition, public notices must be posted outside the offices of both the tribunal and the billing authority.

10.7.3   As far as possible, all facts should be agreed and the areas of dispute identified. The valuation tribunal expects such basic issues as measurements to be agreed, together with the format of valuations and the devaluation of comparables.

10.7.4   Failure to agree is likely to be taken as indicating skimpy negotiations and poor court presentation and, at worst, disrespect for the valuation tribunal, although it does, of course, take two to negotiate!

10.8   Evidence

10.8.1   The following documentation, if presented to the valuation tribunal, is to be accepted as admissible as evidence of fact, unless the contrary is shown (ibid. reg. 41 (2));
(a)   rent returns (see 8.7 and 10.13);
(b)   contents of a rating list (see chapter 8); and
(c)   contents of a completion notice (see 2.4.28–2.4.46 and 9.12).

10.8.2   Evidence may be taken on oath or affirmation (reg. 40 (7) Non Domestic Rating (Alteration of Lists and Appeals) Regulations 1993 (SI 1993 No. 291)). The onus of proof is on the appellant.

10.8.3   The tribunal is required (ibid. reg. 40 (13)) to seek to avoid formality in its proceedings, in so far as it appears to be appropriate, and the normal rules relating to the admissibility of evidence are relaxed for such hearings. This allows the presentation of arguments and evidence by parties who are not legally trained in advocacy and in giving evidence.

10.8.4   Evidence must, however, be susceptible of proof in order to strengthen the credibility of the witness.

10.8.5   The valuation tribunal will accept anything relevant as evidence, but greater reliance will be placed on substantiated evidence (*Robertson Brothers (Brewers) Ltd. v. Houghton & Chester-le-Street Assessment Committee* (1937) and *Garton v. Hunter (VO)* (1969)).

10.8.6   The valuation tribunal is, however, required to conduct the hearing in such a manner as it considers most suitable to the clarification of the issues before it, and, generally, to the just handling of the proceedings reg. 40 (13) 1993 regulations. It seems, therefore, that the rules of natural justice apply.

10.9   Procedure

10.9.1   The valuation officer is never the appellant, since he does not make proposals to alter the list, but he will normally present his case first where it relates to an appeal against his determination that a proposal is invalid and where the appeal relates to an alteration he made to the rating list (ibid. reg. 40 (8) (a)).

10.9.2   With that proviso, the procedure is determined by the president of the tribunal. The decision must be based on evidence and argument presented to the tribunal (rules of natural justice), and, with the exception of the cases listed in 10.9.1, it is usual for the maker of the proposal to present his case first.

10.9.3   The valuation tribunal has a right of inspection of the subject property (ibid. reg. 40 (11)), and the parties to the case are given notice of the inspection and an invitation to attend.

10.9.4 Parties are entitled to examine witnesses and these may be subject to cross-examination by any other party

10.10 Tribunal's decision

10.10.1 At the end of a hearing, the tribunal is required to present its decision orally, although its decision will be presented in writing later. Where there has been no hearing, the tribunal is obliged (ibid. reg. 43 (3)), as soon as is reasonably practicable, to notify the parties in writing of its decision and to give reasons for its decision.

10.10.2 The decisions of valuation tribunals are recorded and are available for public inspection.

10.10.3 The valuation tribunal has no power to award costs against any party and anyone who employs a representative is responsible for the fees of that representative.

10.11 Alteration of the list

10.11.1 The tribunal is empowered (ibid. reg. 44) to require the valuation officer to alter the list (or alter the determination or certification given by him) and he is obliged to comply with the order within two weeks of its making (ibid. as amended by Non Domestic Rating (Alteration of Lists and Appeals) (Amendment) Regulations 1995 (SI 1995 No. 609).

10.11.2 A duty is imposed on the clerk to ensure that the decisions and orders made pursuant to earlier regulations are recorded and that each party to the appeal to which the entry relates is sent a copy of the entry.

10.12 Power to review decisions

10.12.1 As an alternative to the appeal process, reg. 45 (Non Domestic Rating (Alteration of Lists and Appeals) Regulations 1993 (SI 1993 No. 291)) sets up a framework for allowing a tribunal, on written application by a party, to review or to revoke a decision in any particular case, or to vary or set aside its earlier decision, on the grounds that:

(a) the decision is wrong as the result of a clerical error;

(b) a party did not appear and is able to show reasonable cause for not appearing; or

(c) the decision is affected by the decision of the High Court or the Lands Tribunal.

10.12.2 As far as is reasonably practicable, the tribunal appointed to consider an application for a review should consist of the same members as constituted the tribunal which took the decision subject to review.

10.12.3 If, on review, the decision is revoked, the tribunal is to set aside any order made in pursuance of that decision and order a rehearing or redetermination before either the same or a different tribunal.

10.13 Rent returns

10.13.1 In order to fulfill their statutory duties and compile rating lists, valuation officers are able (Sch. 9 para. 5 (1) 1988 Act) to require occupiers or owners to provide them with information regarding the rent paid for hereditaments in their area (see 8.7).

10.13.2 Such information can be used in court proceedings provided the valuation officer gives two weeks' notice to the other parties to the appeal. Such parties are entitled (reg. 41 (3) (Non Domestic Rating (Alteration of Lists and Appeals) Regulations 1993 (SI 1993 No. 291)) to inspect, copy or take extracts from those rent returns.

10.13.3 In addition, any such person notified can require the valuation officer to give him sight of rent returns which relate to hereditaments 'which are comparable in character or otherwise relevant to that person's case' (ibid. reg. 41 (4)). The other party has the right to identify up to four such hereditaments (or as many as those specified by the valuation officer, if that number is greater than four) (ibid. reg. 41 (5)) and to inspect, copy and make extracts of any rent returns held by the valuation officer relating to the hereditaments specified and to require the valuation officer to produce such evidence at the tribunal proceedings.

10.13.4 If the valuation officer does not hold a rent return on any of the hereditaments specified by the other party, there is no right to identify other hereditaments as replacements, and it is clear that if the valuation officer chooses not to rely on the evidence of rent returns, no one else can either.

10.13.5 It seems that rent returns used in court proceedings can only be obtained from hereditaments located within the area covered by the billing authority for whose area the valuation officer is appointed, even if the valuation officer is also appointed for another local authority area.

10.13.6 The purpose for which a valuation officer can acquire such evidence is accepted (e.g. *Smith v. Moore (VO)* (1972)) as being to fulfill his duty to compile and maintain the list. In order to achieve that end, rent returns can be required both to create rateable values for a new list and also to decide whether to alter a list or to disagree with a proposal to alter a list. Rent returns obtained for these purposes can be presented as evidence in related court proceedings (valuation tribunal and Lands Tribunal) (reg. 41 (2) (Non Domestic Rating (Alteration of Lists and Appeals) Regulations 1993 (SI 1993 No. 291)).

10.13.7 However, where the valuation officer obtains a rent return merely to support existing evidence for an adjourned court case hearing, that rent return is inadmissible, because it was obtained specifically for the adjourned hearing and not for the purposes of either creating the rateable value or deciding whether or not to object to a proposal (*Lach v. Williamson (VO)* (1957)).

10.14   Completion notices

10.14.1 Valuation tribunals are required to determine appeals against completion notices (see 9.12), under Sch. 4A, para 4, 1988 Act. A notice of appeal must be sent to the clerk of the valuation tribunal within four weeks of the service of the completion notice (reg. 29 (1) (Non Domestic Rating (Alteration of Lists and Appeals) Regulations 1993 (SI 1993 No. 291)).

10.14.2 Such a notice of appeal must be accompanied by a copy of the completion notice and a statement of the grounds on which the appeal is made (ibid.). The clerk serves a copy of the notice on the relevant authority, within two weeks of its receipt (ibid. reg. 29 (2)).

10.14.3 It is possible (ibid. reg. 45 (1) (6)) to ask for a review of the decision relating to a completion notice in circumstances where new evidence, the existence of which could not have been ascertained by reasonably diligent inquiry, has become available since the conclusion of the proceedings.

10.15   Certificates of value

10.15.1   Under reg. 35, Non Domestic Rating (Chargable Amounts) Regulations 1994, a valuation officer can provide a certificate of value reflecting the rateable value for a hereditament on which transitional relief can be calculated (see 4.6.4–16).

10.15.2   Regulations allow an interested person to appeal against the certification within six months of the provision of the certificate (see 9.11). Appeal is made to the valuation officer and must include the reasons for dissatisfaction (reg. 30 1993 regulations).

10.15.3   The issue can be resolved by the notice of disagreement being withdrawn, by an agreement being reached between the parties or by a decision of the valuation tribunal (ibid.).

10.16   Appeal to Lands Tribunal

10.16.1   The right of appeal against a decision for an order given by a valuation tribunal is to the Lands Tribunal and this includes appeals against completion notices (see 10.14) and certificates of value (see 10.15) (ibid. reg. 47 (1).

10.16.2   Within four weeks of the valuation tribunal's hearing, a written appeal must be made to the registrar to the Lands Tribunal, together with all necessary documentation (ibid. reg. 47 (1)). Appeals are only available to those parties who attended the valuation tribunal hearing or who made written representations (if the case was so determined) or who have made an application for review (ibid. reg. 47 (2)).

10.16.3   The Lands Tribunal has the power to confirm, vary, set aside, revoke or remit the decision or order of the valuation tribunal and may make any order the tribunal could have made (ibid. reg. 47 (6)).

10.16.4   Appeals from the Lands Tribunal on a point of law only are available to the Court of Appeal and thence to the House of Lords.

Any reconsideration of a valuation which stems from the decision on a point of law by the appeal courts is referred back to the Lands Tribunal for determination.

10.17   Arbitration

10.17.1   It is open to the persons who would be parties to an appeal to the Lands Tribunal to agree in writing that the question or issue in dispute is to be referred to arbitration. In these circumstances s. 31 of the Arbitration Act 1950 has effect (ibid. reg. 48 (1)).

10.17.2   In these circumstances, the arbitration award may include any order which could have been made by a tribunal in relation to this question.

10.18   Check-list

10.18.1   Appeals against proposals, etc., can be determined by the valuation tribunal (10.3).

10.18.2   The valuation tribunal normally comprises three lay members who are supported by a professionally experienced clerk (10.4).

10.18.3   Cases can be dealt with by written representation and a pre-hearing review may be used to clarify the issues to be dealt with at the hearing (10.5).

10.18.4   The procedure is determined by the valuation tribunal (10.9).

10.18.5   The valuation tribunal has the power to review its decisions (10.12).

10.18.6   There is a right of appeal to the Lands Tribunal (10.16).

Chapter 11

Rate collection and recovery

11.1 Synopsis

11.1.1 Billing authorities send out rates-demand notices to all ratepayers in their area. Ratepayers are able to pay by instalments if they choose.

11.1.2 If rates are not paid in full, the billing authority can apply to the magistrates' court for a liability order, under which goods can be distrained from the premises of the defaulting ratepayer and sold to cover the outstanding debt.

11.2 Demand notices

11.2.1 Each billing authority is required (reg. 4 (1) and (3) Non Domestic Rating (Collection and Enforcement) (Local Lists) Regulations 1989 (SI 1989 No. 1058)) to serve written demand notices on every ratepayer liable to pay rates to the authority in each financial year.

11.2.2 Demand notices are served by the billing authority on (or as soon as practicable after) 1 April each year in respect of occupied liability for every day during which the hereditament is shown in the rating list and occupied by the ratepayer (ibid. and s. 43 (1) 1988 Act). In respect of unoccupied liability, demand notices are served by the billing authority for every day during which the hereditament is not occupied, the ratepayer is the owner, the hereditament is shown in the rating list and no regulation exempts the ratepayer from liability (ibid. reg. 4 (1) and (3) and s. 45 (1) 1988 Act).

11.2.3   Demand notices may be served before the beginning of any financial year provided that the level of Uniform Business Rate has been fixed and that the conditions regarding the status of the ratepayer and the hereditament (reg. 5 (2) Non Domestic Rating (Collection and Enforcement) (Local Lists) Regulations 1989 (SI 1989 No. 1058)) (see 11.2.2 above) are likely to be fulfilled.

11.2.4   Demand notices must contain those matters specified in the relevant regulations (Council Tax and Non Domestic Rating (Demand Notices) (England) Regulations 1993 (SI 1993 No. 191); Council Tax and Non Domestic Rating (Demand Notices) (Wales) Regulations 1993 (SI 1993 No. 252) and Council Tax and Non Domestic Rating (Demand Notices) (City of London) Regulations 1993 (SI 1993 No.149), as amended). Demand notices must, therefore, state the address and description of the hereditament, and its rateable value and be accompanied by explanatory notes, including expenditure estimates.

11.2.5   Provision is made for the payment of rates by instalments within Schedule 1 of the Non Domestic Rating (Collection and Enforcement) (Local Lists) Regulations 1989 (SI 1989 No. 1058), although not more than ten annual instalments are permitted.

11.3   Non-payment of rates

11.3.1   Where a ratepayer fails to pay rates in accordance with the instalments required by a demand notice, the billing authority serves a further notice identifying the instalments to be paid. If these remain unpaid, the amount outstanding becomes payable within seven days.

11.4   Enforcement

11.4.1   Rates due are recoverable by distress, liability order or by civil debt. In certain cases, a failure to pay rates can result in a commitment to prison. Debts due to ratepayers from billing authorities are also recoverable by civil debt.

11.4.2   *Distress*
Distress, or the distraining of goods, is a process whereby goods up to the value of the debt are taken from the debtor's premises

and sold. The proceeds are used to pay off the debt (including the cost of the distress) and any surplus is paid back to the debtor. Distress is the traditional remedy for rate recovery and is normally exercised by bailiffs with a liability order (see 11.4.4–7) from a magistrates' court.

11.4.3   Prior to such action, the billing authority is required (ibid. reg. 11 (1)) to serve a reminder notice on the ratepayer stating the amount owed. Seven days after such a notice is served, the billing authority may apply to the magistrates' court for a liability order against the ratepayer (ibid. reg. 12 (1)).

11.4.4   ***Liability Order***
The application for a liability order is made in the form of a complaint to a Justice of the Peace requesting the issue of a summons directing the ratepayer to appear before the court to explain why the amount outstanding has not been paid (ibid. reg. 12 (2)). The court may proceed provided that it is satisfied that the summons was served on the ratepayer and may hear the complaint in his absence.

11.4.5   The duty of the court (ibid. reg. 12 (5)) is to make the liability order if it is satisfied that the sum has become payable by the ratepayer and has not been paid. The order is made to cover the amount of rates owing to the billing authority plus the authority's costs in obtaining the order (ibid. 12 (6)).

11.4.6   Once the liability order has been made, the billing authority may levy the appropriate amount by distress and proceed with the sale of goods belonging to the ratepayer. Basic domestic items cannot be seized and sold for this purpose (ibid. reg. 14 (1A)), and if the amount owing is paid or offered, the authority must accept it and not proceed with the distress.

11.4.7   Anyone implementing the liability order and distraining on the goods of a ratepayer must have written authorisation from the billing authority and leave with the ratepayer a copy of the relevant regulations and a memorandum setting out the amount due.

11.4.8   There is a right of appeal to the magistrates' court by any person who is aggrieved by the levying of distress (ibid. reg. 15 (1)). The appeal is effected by making a complaint to a Justice of the Peace and requesting the issue of a summons directed to the authority in question to appear before the court to answer the matter by which the person is aggrieved.

11.4.9   If satisfied that the levy was irregular, the court can order the authority to discharge the goods distrained and to award compensation for goods distrained and sold in the amount which, in the opinion of the court, is equal to the amount which would have been awarded by special damages in an action for trespass (under the procedures in reg. 15 (3) ibid.).

11.4.10 *Commitment to prison*
Where the ratepayer against whom the billing authority has attempted to levy distress is an individual and there were insufficient goods to cover the amount of the debt, the billing authority can apply (ibid. reg. 16 (1)) to the magistrates' court for the issue of a warrant committing the debtor to prison. The court must inquire, in the presence of the debtor, what his means are and whether the failure to pay was due to willful refusal or culpable neglect. Only if, in the court's opinion, the debtor's failure to pay was due to wilful refusal or culpable neglect will the court issue the warrant of commitment (i.e. a coercive means to force payment) or fix a term of imprisonment and postpone the issue of the warrant until such time and on such terms as it thinks fit (reg. 16 (2)(3) ibid.).

11.4.11 The court has the power to remit all or part of the debt (rates unpaid and billing authority's costs) following inquiries into the debtor's means described in 11.4.10 (ibid. reg. 17 (2)).

11.4.12 *Insolvency*
Where a liability order has been made against a debtor, the amount due is treated as a debt for the purposes of the Insolvency Act 1986 (ibid. 18 (1)).

11.4.13 Once a warrant of commitment has been issued against a debtor, or a term of imprisonment fixed, no further steps may be taken against him by way of distress or bankruptcy or winding up (ibid. 19 (1)). While steps are being taken against a debtor by way of any one of these remedies, no other method of enforcing the debt can be pursued (ibid. 19 (2)). Distress, however, can be resorted to more than once (ibid. 19 (3)).

11.4.14 *Appeal to the High Court*
Any person who was a party to proceedings before a magistrates' court or who is aggrieved by its decision may question the proceedings by case stated on the grounds that it is wrong in law or is in excess of the court's jurisdiction (s. 111 Magistrates' Court Act 1980).

11.5 Civil debt

11.5.1 Where no liability order has been made, a billing authority can recover unpaid rates 'in a court of competent jurisdiction' (reg. 20 (1) Non Domestic Rating (Collection and Enforcement) (Local Lists) Regulations 1989). Once such proceedings have been initiated, no liability order can be made (ibid. reg. 20 (2)).

11.5.2 This action is available to a ratepayer to whom money is owed by a billing authority, i.e., where overpaid rates have not been repaid to the ratepayer (ibid. reg. 22).

11.6 Check-list

11.6.1 Billing authorities serve demand notices for rates payable on occupiers or owners of hereditaments (11.2).

11.6.2 Where rates demanded are not paid following the service of a reminder notice, payment may be enforced by: distress (11.4.2–11.4.9); liability order (11.4.4–11.4.9); or civil debt (11.5).

11.6.3 There is an appeal to the High Court against the decision made by a magistrates' court.

Chapter 12

Criticisms of the rating system

12.1 Synopsis

12.1.1 The Uniform Business Rate in the UK is a tax levied by central government on the value of non-domestic property and paid by the occupier (or, where there is no occupier, the owner) raising money for local authority expenditure. It is, therefore, assigned revenue.

12.1.2 Like all taxes, the UBR should conform to the recognised and socially accepted taxation principles of equity.

12.1.3 The UBR should be investigated to ensure that it is a fair and socially acceptable system and that it is implemented equitably.

12.1.4 Where there are inadequacies, unfairness or public dissatisfaction, the system should be reformed.

12.2 Generally

12.2.1 Criticisms can be made against the Uniform Business Rate at various levels. It is possible to criticise the principles of the tax and its implementation. It is also possible to criticise its very essence as a tax on the value of the occupation of land, i.e. a necessity. It is, after all, a tax and: 'To tax and to please ... is not given to men' (Edmund Burke, *Speech on American Taxation*, 1774).

12.2.2 The purpose of this Chapter is to provide an overview of some of the problems associated with the UBR as a land tax within the UK and to encourage further debate. Readers should refer to the Bayliss Report (The National Committee on Rating. *Improving the*

System. The Royal Institution of Chartered Surveyors, 1996), the conclusions of which appear as Appendix C, to the Layfield Report (*The Report of the Committee of Inquiry on Local Government Finance,* HMSO, 1976) and to other relevant texts, some of which are listed in the bibliography.

12.3   Taxation of land

12.3.1   The opportunity to tax land is one which is taken by almost every administration in the world. In some cases, the tax is levied by the central government, e.g., in China, and in the UK with the UBR; and in other cases the tax is levied by states within a federal system, e.g., the United States and Malaysia, and in other cases, the tax is levied by the municipalities, e.g., France and Cyprus.

12.3.2   There is a fundamental conflict within any taxation system and that is the conflict between the need for the governments to raise funds out of which to provide those services which are required by their citizens and the desire of those citizens to minimise their liability to pay taxes.

12.3.3   This conflict is ideally remedied (or at least reduced to bearable proportions) by imposing a variety of taxes within a rigid, efficient and effectively implemented legal framework, so that there is minimal tax evasion and all those who are required to pay taxes do so and can be seen to do so. This ensures that all who are in a similar situation are treated in the same way (i.e. 'horizontal equity') and also provides the taxpayer with an element of choice in how disposable income is spent.

12.3.4   There is, for example, a tax in the UK on the purchase of petrol and a licence fee paid for the taxing of a motor vehicle. Those who choose to drive a motor vehicle are, implicitly, volunteering to pay those taxes. Those who choose not to drive, do not pay those taxes. This element of choice exists in the payment of other taxes (not all of them payable on the purchase of items) and allows taxpayers the chance to arrange their affairs so that they minimise the tax they pay. (This is the principle of tax avoidance, which is legal in the UK. Tax evasion, which is deliberately failing to pay taxes due, is not.)

12.3.5   Nevertheless, if all citizens chose not to pay 'avoidable' taxes, the government would have no funds out of which to provide necessary services. It is, therefore, to ensure some certain and predictable revenue, that an unavoidable tax is implemented, such as that levied on the occupation of land and buildings.

12.3.6 Land is a necessity. There is no physical human activity which can be undertaken in the UK without the use of 'land' (defined in its broadest sense). The taxing of land is, therefore, the taxing of a necessity and individuals are given no choice about whether or not to become taxpayers. (There may, however, be an element of choice as to the exact nature of the land and buildings occupied and therefore the amount of tax payable.)

12.3.7 This removal of the element of choice is one reason why land taxes cause so much concern, both to taxpayers and to other interested parties.

12.3.8 It is, however, axiomatic that if taxes are to be paid, they should be paid by those with 'wealth' (however that is defined). It can be argued that, because taxes are paid in cash, the 'wealth' of a taxpayer should be calculated on the basis of funds available to pay taxes, i.e., on the basis of ability to pay, which is normally expressed in terms of the level of disposable income received. But to limit taxation to an income tax alone would leave significant amounts of 'wealth' (in the form of capital assets) untaxed, and it is generally considered that a balance should be struck between the various kinds of 'wealth' and the degree to which each is taxed.

12.3.9 In the UK, income is subject to income tax, and there are various other capital taxes, e.g., Capital Gains Tax and Inheritance Tax. However, these capital taxes are levied only on the occurrence of specific and, often, rare events (normally a sale of the asset and a death, respectively).

12.3.10 The ownership or occupation of landed property is generally recognised in the UK as a sign of 'wealth', like the possession of such valuables as paintings or jewelry. All these are taxable on disposal through the Capital Gains Tax and Inheritance Tax legislations and, while land and buildings too are liable to such taxes on disposal, there are practical reasons of liability why such assets as paintings and jewelry cannot be taxed on a more comprehensive or regular basis.

12.3.11 The ownership or occupation of landed property is, however, a visible and publicly assessable sign of 'wealth', the value of which can be attributed, at least in part, to the benefits derived from the local, regional and national environment in which it is located. The right of a taxing authority to take back some of that attributed value (or 'betterment') in the form of a property tax is part of the justification for taxing land.

Without community activity, land has a bare subsistence value only. Any value above subsistence has been created by the community in providing roads, sewers, public utilities and the demand for housing, factories and the like. It is therefore pre-eminently fair and sensible that that increase in value should be payable back to the community in the form of a tax ... (Hector Wilks. (1985) *A case for the present system.* 275 *EG* 26 & 28, at p. 26).

12.3.12 It can be argued that the taxation of land (and buildings constructed on that land) should only be used as a means of taking back from the taxpayer that amount of annual value which the taxpayer has not created, i.e., it is necessary to assess the value which the 'community' (defined to include local, regional and national agencies) has contributed to the value of the property over and above any contribution by the owner. This increased value or betterment is the amount which should be liable to tax and would give the community in its various tiers of government a fund from which to continue improving the environment, to the benefit of all.

12.3.13 The corollary is, of course, that when the community depreciates the value of someone's property, compensation is payable to reflect that depreciation, and there are few arguments against this principle.

12.3.14 However, if taxation of land is to be used as a means of drawing back betterment into the hands of the community, then certain principles should be observed. For example:
(a) it should be presented as a 'betterment tax' to the taxpayers and to the nation at large (such systems were introduced in 1967 (Land Commission Act 1967) and in 1975 and 1976 (the Community Land Act and the Development Land Tax respectively));
(b) it should affect all land (there is an argument for excluding the value of the buildings which are the creation of the owner, normally, of course with the benefit of planning permission, however);
(c) it should tax no more than the increase in value resulting from the activities of the 'community' and, ideally,
(d) the revenue should be 'ring-fenced' for community projects and for the payment of compensation on depreciation in the value of privately-owned land, caused by the 'community' in such projects as the construction of motorways. (See, for example, Hector Wilks. 'Property tax systems'. *The Valuer.* January/February 1987.)

12.3.15   The Uniform Business Rate is clearly not such a betterment tax, although (with the exception of Capital Gains Tax and the set-off provisions of section 7 of the Compulsory Purchase Act 1965), it is the only national legislation within the UK which makes any attempt to tax the betterment value of land. Note, however, that the value taxed by the UBR is not exclusively betterment value, because the intrinsic value of land and the value added by the owner or occupier would be included within the market value and therefore within the rateable value of the hereditament. Capital Gains Tax and the set-off provisions in s. 7 of the Compulsory Purchase Act 1965 are, however, one-off payments and are extremely selective in their application. The UBR is not so selective in its application and is levied on an annual basis.

12.3.16   While the UBR can be justified on the basis of being (at least in part) a tax on betterment, it must be criticised because of its limited application. To be capable of being justified within this context, the UBR must be universally applied and there are two major property types which are excluded from the UBR, namely domestic property and agricultural land and buildings.

12.3.17   The UBR has been criticised as being a tax on improvements in the same way that it is criticised as being a tax on a necessity, though it is hard to see how else the tax could be implemented with any degree of equity. (See Appendix C The Bayliss Report – RICS, 1996, para. 3.10 on p. 15 and 16.)

12.3.18   The Uniform Business Rate exists in its present form, however, because of the reform of the pre-1990 rating system (see Appendix A), and is, therefore, something of a political solution to a series of problems rather than a serious attempt to deal with the issue of land taxation in the UK. This is not to excuse its failings, but to set them into a context for further debate.

12.4   Principles of the tax

12.4.1   The Uniform Business Rate was introduced in 1990 following many years of criticisms of and investigation into the pre-1990 rating system (see, for example, *The Report of the Committee of Inquiry on Local Government Finance* (The Layfield Report) HMSO, 1976). It taxes only non-domestic property (see Chapters 3 and 4) and the level of tax imposed is increased annually by no more than the rate of inflation (see Chapter 1). It is a tax levied, notionally, by central government, but its revenue is given to local government to spend, i.e., it is an assigned revenue (see Chapter 1).

12.4.2   The rationale behind the assignment of the tax by central government to local government seems to be that, while local government requires the revenue from local businesses to pay for its services (at least in part), local government cannot be trusted to levy a rate which minimises local government expenditure and allows local businesses the chance to budget ahead for its occupational costs. (This is evidenced, too, in the way the level of the Council Tax which is fixed by local government is liable to be 'capped' by central government so that local authorities are financially penalised if they attempt to increase the level of Council Tax levied in their area over a certain level – see 13.8).

12.4.3   The fixing of the UBR by central government (as opposed to local government) can be justified on the basis of electoral representation. All adults resident in the UK (with UK and Republic of Ireland citizenship) have the right to vote in national elections and those who are liable also pay central government taxes. They, therefore, have a right, by means of the ballot box, to influence the nature of central government taxation. Within local government, the right to vote exists only for local residents and there is no right to influence the nature of local government taxation by means of the ballot box for owners of local businesses, unless the proprietor also happens to be a local resident. Before 1990, local government was required to take into account the views of local commercial interests in fixing the level of taxation, but this was considered to be an inadequate and ineffective system for ensuring that local businesses were not penalised by excessive and disparate tax rates throughout the country.

12.4.4   The removal of the responsibility for fixing the level of the tax from local government has had several effects. For example, local government has lost direct control over about 20% of its income (see Table 1.2), thereby increasing the level of its financial dependence on central government to about 84%. The reduction in financial independence has meant, inevitably, a reduction in the freedom of local authorities to vary the range and level of the services they provide.

12.4.5   This further reduces the level of local democratic accountability which was a stated aim of central government when introducing both the Community Charge (or poll tax) and its replacement, the Council Tax (see Chapter 15 and Appendix A).

12.4.6   The principle of local democratic accountability is one of the arguments for returning non-domestic rates to local authorities and for treating domestic properties in exactly the same way as non-domestic property, i.e., for abolishing the Council Tax and for making domestic property liable to the UBR. This would give local authorities control over about 35% of their sources of finance. The argument against this is considered to be largely political.

12.4.7   The UBR can be criticised because, like the pre-1990 rating system, it suffers from a number of illogical exemptions, such as the exemption from the UBR for agricultural land and buildings. The removal of the exemption for Crown occupations in the Local Government and Rating Act 1997 is a long-awaited reform which will remove one major criticism against the implementation of rating principles. In the Bayliss Report (The National Committee on Rating. *Improving the System*. The Royal Institution of Chartered Surveyors, 1996), which recommended the establishment of a Committee to review the current exemptions, the issue was summarised as follows:

> Exemption from rates has been a matter of some controversy over the years. It is now generally accepted that certain public facilities such as places of public religious worship and public parks are properly exempt, but exemption for many others is now widely seen more as a matter of expediency than deserving. (ibid. para. 6.2.1 at p. 32 and see also Appendix C)

12.4.8   A system of land taxation can only be perceived as fair if all land is subject to taxation. If it is considered politically expedient to exempt a particular occupier, e.g., because of poverty or because it is considered to be socially acceptable to do so (as could be argued for places of public religious worship), then the tax liability of such occupiers should be undertaken by central government (see, for example, *The Report of the Committee of Inquiry on Local Government Finance* (The Layfield Report) HMSO, 1976, para. 63 at p. 168).

12.4.9   Another principle of the tax for which it can be criticised is that the rate is presented as a tax on the occupier and the occupier is perceived to bear the burden of the levy while enjoying few of the long-term advantages of the capital value of the property which the community has created (see 12.3.8–12).

12.4.10   It can be argued that the incidence of the UBR is, in fact, an indirect burden on the landlord. It is well recognised (refer equation theory, 6.4.24–6) that, because of the rate burden, less rent is payable by an occupying tenant. Therefore, the UBR should be levied directly on the landlord, who would require a contribution to that liability from an occupying tenant, who would, thus, end up paying the same level of occupational costs (rent and rates) as if the tax were levied directly on the occupier, and the burden and incidence of the tax would be clearly visible as falling on the owner. There would be few practical difficulties in identifying an owner to pay the rates and the tax would be and would be seen to be an impost on the owner.

12.4.11 There is considered to be a high level of ignorance on the part of ratepayers regarding the UBR system itself and the role of the various bodies responsible for its operation (The National Committee on Rating. *Improving the System*. (The Bayliss Report) The Royal Institution of Chartered Surveyors, 1996, and *The Report of the Committee of Inquiry on Local Government Finance* (The Layfield Report) HMSO, 1976). Whilst this is not necessarily an intrinsic fault of the system itself, it does make public acceptance of the tax less likely and, therefore, its administration harder and more expensive.

12.4.12 The system in Scotland is under a different jurisdiction to that in England and Wales. Each jurisdiction has its own method of operations and statutory timetables (for example, there is no central rating list in Scotland) and there are significant differences in the functions and financing of local government in Northern Ireland. It is a recommendation of the Bayliss Report (The National Committee on Rating. *Improving the System*. The Royal Institution of Chartered Surveyors, 1996), which recommended (ibid. para. 1.3.2 at p. 2) that the taxing provisions be consolidated and harmonised throughout the UK.

12.5   Implementation

12.5.1 The social acceptability of the UBR could be dramatically improved by the removal of the transitional arrangements which currently operate to ensure that, on both the 1990 and 1995 revaluations, the increases in rates which were phased in for certain occupiers were paid for by the phasing in of the decreases in rates for those occupiers entitled to a reduced rate liability.

12.5.2 Transitional arrangements immediately obviate the implied fairness of a revaluation and cannot be justified on any grounds other than the political expediency of ensuring that transitional relief is self-financing. If it is desirable from the political standpoint to phase in increased rate liability for certain occupiers, then any deficit to the rate revenue as a result of transitional relief should be paid for by central government and not by those ratepayers who, by definition, have been paying and are forced to continue to pay more than their liability under the strict rules of the phasing system.

12.5.3   The use of formulae to assess a rateable value for the operational hereditaments occupied by the so-called statutory undertakers of such enterprises as electricity, gas, telecommunications and water cannot ensure that such occupiers pay the same proportion of their rate liability as any other occupier. Central government has stated its intention to return such hereditaments to conventional methods of valuation to ensure that each pays UBR on the same basis as all other tax-payers, and it is expected that such provisions will be implemented for the next revaluation in 2000.

12.5.4   There is, currently, an anomaly within the relief given for empty property. Owners of empty commercial (non-industrial) hereditaments are liable to pay the non-occupied rate, which is 50% of that payable by an occupier, while owners of empty industrial and warehouse premises enjoy a relief of 100% of the occupied rate. This is clearly an unjustifiable situation at a time when recession in the UK economy does not distinguish between the industrial and the non-industrial sectors. The relief given to the industrial occupiers dates from 1984 and 1985 when, as a result of the then industrial recession, 100% rate relief was given to occupiers of empty 'industrial and storage hereditaments'. This relief continued at a time when other parts of the UK economy were suffering the effects of an economic recession. Such discriminatory treatment of occupiers of different hereditaments is not justified and should be abandoned.

12.6   Conclusion

12.6.1   The success of any tax can be measured in a variety of ways, e.g. cost-efficiency, certainty and predictability of yield, and social acceptability (see Appendix F). It is, perhaps, within the criterion of social acceptability that the fundamental success of any tax should be measured.

12.6.2   Part of achieving such success is in the presentation of the tax to the nation as a whole, so that it is seen to reflect the priorities, standards and aspirations of the nation as well as to be responsive to any relevant social changes.

12.6.3   It is also true that the structure of local authority finance must be compatible with (and ideally should be reformed alongside) the structure of local authorities themselves together with their responsibilities.

12.6.4   To date, these are not principles which have been recognised within the UK. The current trend of tinkering with the system to make it more palatable cannot be expected to deal with the fundamental problems of the system. Indeed, such tinkering makes matters worse by increasing the already impressive number of legislative documents which relate to an already complicated taxation system.

12.6.5   The problems associated with the 1990–93 reforms of local authority revenue (see Appendix A) mean that there is little political will to revisit this issue so soon after the events and there is not likely to be a fundamental reform of local authority finance in the foreseeable future.

Chapter 13

Council tax

13.1   Synopsis

13.1.1   The Council Tax replaced the Community Charge (poll tax) with effect from 1 April 1993.

13.1.2   The Council Tax is levied on owners and occupiers of domestic property by local (billing) authorities.

13.1.3   Half the tax payable is a personal element, which assumes that there are two taxable adults resident, and the other half is a property element, which is based on the banded value of the property.

13.1.4   The level of the Council Tax is fixed by the local authorities, subject to the capping powers of central government.

13.2   Introduction

13.2.1   The Local Government Finance Act 1992 (LGFA 1992) provides for certain local authorities to levy and collect the Council Tax. Details of the implementation of the Council Tax can be found in the relevant Council Tax Practice Notes (see Bibliography).

13.2.2   The Council Tax has two elements: a personal element which makes up 50% of the bill and a property element which makes up 50% of the bill. The details of the Personal element are explained in this Chapter (see 13.4), while the property element is explained in Chapter 14.

13.2.3   The Council Tax is therefore something of a hybrid tax based in part on the pre-1990 rating system, in that it is a tax on the value of the dwelling, and in part on the 1990–93 Community Charge (poll tax) in that liability reflects the number of residents.

13.3   Liability

13.3.1 Liability for the Council Tax was not clearly established in the consultation document (HMSO, 1991). It stated (p. 14) that:

a single bill should be sent to each household. It would be for the household to decide how the bill should be divided between those living in the property and to make the appropriate financial arrangements between themselves.

13.3.2 The document continues that:

The old rating system worked in this way. These arrangements would be appropriate for most households.

13.3.3 However, the 'old rating system' was a tax levied on occupiers, the definition of which was established by considerable amounts of case law (see 2.3). The implication of the consultation document is, therefore, that the Council Tax is an occupiers' tax, which is reinforced in the Summary to the consultation document (ibid. page unnumbered, item 7) where it is stated that:

The liability for the new council tax will fall on the occupier, normally the head of the household.

13.3.4 *Hierarchy of tenure*
However, within the enacting legislation, residents of dwellings are liable (s. 6 of the LGFA 1992) according to a hierarchy of their interests in the property.

13.3.5 The first priority is given to residents with a legal interest, that is:
(a)   residents with a freehold interest;
(b)   residents with a leasehold interest which is not inferior to another;
(c)   residents who are statutory or secure tenants;
(d)   residents with a contractual licence to occupy.

13.3.6 Next, there are residents with no legal interest. This appears to refer, particularly, to squatters and trespassers. It may also apply to charitable and service occupiers, although such residents may fall under the previous category if they have a contractual licence.

13.3.7   Finally, there is the owner, who is defined (s. 6 (5) LGFA 1992) as the person with a material interest in the whole or any part of the dwelling where at least part of the dwelling is not subject to a material interest inferior to his interest. 'Material interest' means a freehold interest or a leasehold interest which was granted for a term of six months or more (ibid. s. 6 (6)). Thus, the owner is only liable where no one else has their sole or main residence in the property. The Secretary of State can also prescribe classes of property which may be the subject of a determination to make the owner liable (see 13.3.11).

13.3.8   The Act identifies (s 6 (5)) a 'resident' as

an individual who has attained the age of 18 years and has his sole or main residence in the dwelling.

13.3.9   There is, however, no definition of 'sole or main residence', although recent case law (*Bradford Metropolitan City Council v. Anderton* (1991)), described a residence as a settled and usual abode and argued that residence should not be defined only by time spent there. The case also established that a ship is not a residence. Liability appeals go to the valuation tribunal (see Chapter 10).

13.3.10   *Students*
Students are exempt from Council Tax when living in halls of residence and hostels, but discounts only apply where students are solely or mainly resident in a chargeable dwelling with or without a resident who is not a student. Students are, therefore, 'disregards' for the purpose of calculating the number of taxable residents.

13.3.11   *Houses in multiple occupation, hostels etc.*
Where it is difficult to levy the Council Tax because there is no single household, e.g., where houses are in multiple occupation or are occupied as hostels, etc., s. 8 (1) 1992 Act and the Council Tax (Liability for Owners) Regulations 1992 (SI 1992 No. 551) specifies classes of dwellings for which the person liable for the Council Tax is the owner, rather than the occupier, as follows:
Class A nursing homes and other similar homes;
Class B houses of religious communities;
Class C houses in multiple occupation;
Class D residences of staff who live in houses occasionally occupied by an employer;
Class E residences of ministers of religion.

Liability for dwellings owned by Ministers of the Church of England is transferred to the Diocesan Board of Finance, rather than to the owner.

13.3.12 *Joint and several liability*
Joint and several liability applies to married couples and to couples living as if they were married and where there are two or more people with identical 'interests', i.e. with interests falling within the same level of the hierarchy of liability (s. 9, 1992 Act).

13.3.13 Further details can be found in *The Council Tax: Practice Note No. 2 – Liability, discounts and exemptions*, (revised July 1993) (prepared by the Department of the Environment, the Welsh Office, and associations of local authorities).

13.4   Personal element

13.4.1 The personal element of the Council Tax accounts for 50% of the bill. Personal discounts are given to reflect single-person households and unoccupied dwellings and apply only to the personal element of the tax, i.e., to the 50% of the tax which relates to residents, and not to the 50% of the tax which relates to the property.

13.4.2 The Council Tax takes into account the number of adults in each household by assuming a basic number of two adults per household.

13.4.3 The most common number of adults per household is two (HMSO, 1991, p. 5) and it was considered (ibid.) that by specifying two as the assumed number of adults, the number of households which would need discounts would be minimised, and that only a small minority of the population would not be taken into account, since the number of people who are the third and subsequent adults living in households is only about four million (ibid. p. 6).

Thus, households with two or more adults pay the full amount of Council Tax for their property.

13.4.4 *Single-person Households*
Single-adult households benefit from a personal discount of 25% of the basic bill (s. 11, 1992 Act).

13.4.5 Thus, discounts on the personal element of the Council Tax are as follows:

(a) where a dwelling is occupied by a single resident, there is a discount of 25% on the full charge for the property, i.e., a 50% discount on the personal element of the charge;

(b) where the dwelling is occupied by two or more residents and all but one qualify as 'disregarded' people (see 13.4.6), there is also a discount of 25% on the full charge for the property (again, a 50% discount on the personal element of the charge);

(c) certain dwellings are excluded completely from the charge, so, strictly speaking, this is an exemption, not a discount. Such dwellings include students' halls of residence, student hostels and dwellings where all the residents are students (see 13.3.10).

13.4.6 Schedule 1 of the 1992 Act (as amended by the Council Tax (Discount Disregards) Order 1992 (SI 1992 No. 548)) provides that the following are 'disregarded' people who qualify for a personal discount under s. 11 of the Local Government Finance Act 1992:

(a) a person in detention, subject to certain conditions;

(b) a person who is severely mentally impaired and who is entitled to one of a series of specified benefits;

(c) a person who is 18 years old and in respect of whom child benefit is payable;

(d) students, student nurses, apprentices and youth-training trainees;

(e) patients within hospitals and hostels, as defined; and

(f) care workers.

13.4.7 The Secretary of State retains the power to extend the list of 'disregarded' people, as appropriate.

13.4.8 The various categories of dwellings giving rise to different liabilities can be summarised as follows:

(a) residentially-occupied dwellings which are someone's sole or main residence – full liability falls on the resident with the prime interest;

(b) empty and unused – no 'resident', so the owner is liable. Such properties attract a discount of the 50% on personal element;

(c) empty and unused dwellings, which are exempt (see 14.7), so there is no liability;

(d) furnished and unused – the owner is liable. Such properties attract a discount of 50% on the bill;

(e) furnished and in use, but not anyone's sole or main residence (e.g. a second home), the discount is 50% of the bill. In Wales, the

Council Tax (Prescribed Class of Dwellings) (Wales) Regulations 1992 (SI 1992 No. 3023) specifies that only one personal discount applies to every dwelling in Wales:
(i) in respect of which a period of six months or more has elapsed   since it was anyone's sole or main residence;
(ii) which is furnished; and
(iii) which is not either a pitch occupied by a caravan; a mooring occupied by a boat; an unoccupied dwelling in respect of which a person qualifies as a personal representative; or a dwelling which is unoccupied because the owner resides elsewhere in job-related accommodation.

13.4.9   Any resident aggrieved by the decision of the local authority on an application for a discount may ask the local authority to review its decision and, if still dissatisfied, may appeal to the valuation tribunal.

13.4.10   *Verification*
Local authorities take their own steps to verify the information provided. There are penalties for making fraudulent use of the discount arrangements.

13.5   People on low incomes

13.5.1   It is the stated intention of central government (HMSO, 1991, p. 17) that

people on low incomes are not called on to make a disproportionate contribution to local taxation.

13.5.2   A system of council tax benefits has been incorporated into amendments to the social security acts.

13.5.3   Local authorities administer the scheme and the subsidies for doing so are broadly in line with those for the Community Charge benefits scheme.

13.5.4   For individuals or couples on income support or equivalent levels of income, rebates meet 100% of their liability under the Council Tax. For further information, reference should be made to *The Council Tax: Practice Note No. 3 – Council Tax Benefit* (prepared by the Department of the Environment, the Welsh Office, and associations of local authorities).

13.6   Local authorities' discretion to fix tax

13.6.1 It was intended (HMSO, 1991, p. 12) that local authorities would retain the discretion to set their budgets above or below the standard level, subject to the restraints imposed by central government which operate at the time.

13.6.2 Central government's intention is that an authority's spending above or below the standard level should be reflected in its Council Tax bills, so that Council Tax payers benefit from efficiency savings and face additional costs for any extra spending.

13.6.3 In this way, central government's intention that the level of Council Tax relate clearly to the spending decisions of the local authority and that the local authority become directly accountable to its taxpayers should be achieved.

13.6.4 Any variation in spending falls proportionately on the bills of all Council Tax payers in the area. This means that each household in an authority's area should see a percentage increase or reduction in their bill equal to that of the level of spending, compared with the standard level.

13.6.5 *Tiers of authorities*
In areas with more than one authority, the tiers should each set their own separate element of the Council Tax, and this should be clearly identified on the single bill sent to the Council Tax payer.

13.6.6 For further information on this matter, refer to *The Council Tax: Practice Note No. 7 – Tax setting, precepting and levying* (revised August 1993 and August 1995) (prepared by the Department of the Environment, the Welsh Office, and associations of local authorities).

13.7   Grants system

13.7.1 Under the Council Tax, the total amount available from the general grant is distributed so that the authorities in each area can finance spending at a standard level by levying the standard taxable amounts fixed by the Secretary of State for the Council Tax, taking into account the personal discounts.

13.7.2 In this way, local authorities are compensated for differences in their expenditure needs, and for any variations in their taxable capacity.

13.7.3   Variations in spending needs are taken into account in the Standard Spending Assessment which represents the amount of revenue expenditure which it is appropriate for each authority to incur to provide a standard level of services.

13.8   Capping

13.8.1   Central government recognises (HMSO, 1991, p. 25) its duty to protect local taxpayers from unacceptably high bills as well as to control the level of public expenditure. The new system incorporates arrangements to ensure restraint in local spending and taxation.

13.8.2   Central government, therefore, introduced capping powers appropriate to the Council Tax.

13.9   Phasing arrangements

13.9.1   Central government's aim (HMSO, 1991, p. 27) was to ensure that no household faced an unreasonable increase in their overall bill from one year to another as a result of the introduction of the Council Tax.

13.9.2   There were requirements on local authorities to provide relief for individual households for this purpose, the cost of which was to be found within the overall total available for central grants to local authorities.

13.9.3   For further details of such phasing arrangements, see *The Council Tax: Practice Note 4 Transitional Arrangements* (as revised) (See References and Bibliography).

13.10   Appeals

13.10.1   Valuation tribunals have jurisdiction to hear appeals relating to liability to pay the Council Tax; the designation of a dwelling as a chargeable dwelling; the calculation of the tax payable (under s.

16 of the 1992 Act); and the imposition of a penalty for failure to supply information about the liable person (para. 3 (1) of Sch. 3 of the 1992 Act).

13.10.2 There are also rights of appeal to the valuation tribunal in respect of valuation matters (see Chapter 14).

13.10.3 There is no right to appeal to the valuation tribunal against the setting of the amount of Council Tax; the specification of a class of exempt dwelling; the billing authorities' determination about the application of a discount to a class of dwelling (in Wales); or the issue of a precept or the allocation of grant or the capping of expenditure by the Secretary of State. These matters may be questioned only on judicial review.

13.10.4 There is no right of appeal to the Lands Tribunal from the valuation tribunal. There is a right to appeal to the High Court on a point of law only (reg. 32 of the Council Tax (Alteration of Lists and Appeals) Regulations 1993 (SI 1993 No. 290) and reg. 32 (1) of the Valuation and Community Charge Tribunals Regulations 1989 (SI 1989 No. 439) (as amended)).

13.11   Billing and collection

13.11.1 Local authorities are given the freedom to design their own Council Tax bills and literature. However, while it is sufficient to address the bill to the 'occupier', a named person is required for recovery methods.

13.11.2 The Council Tax bill must also state:
(a)   the amount of tax for up to three precepting authorities; and
(b)   any discounts, transitional relief, lump-sum allowance and benefits, as applicable.

13.11.3 In addition, the authority is required to state the assumptions made in calculating the bill. This can be done on the bill itself.

13.11.4 Regulations allow for the payment of the tax by instalments. Further details can be found in the *Council Tax: Practice Note No. 5 – Administration (including billing and collection)* (revised April 1994) (prepared by the Department of the Environment, the Welsh Office, and associations of local authorities).

13.12   Check-list

13.12.1 The Council Tax comprises two elements: a personal element, worth 50% of the bill and a property element, worth 50% of the bill (13.2.2).

13.12.2 Liability for the Council Tax rests initially with the legal owner for whom the dwelling is the sole or main residence. There is a hierarchy of tenure which includes a resident and joint and several liability (13.3).

13.12.3 Owners are liable for the Council Tax for certain classes of dwelling, e.g. houses in multiple occupation (13.3.11).

13.12.4 There is a personal discount of 25% of the bill for dwellings with only one resident or where only one resident is not 'disregarded' (13.4).

13.12.5 There is a rebate of up to 100% of the bill for occupiers on low incomes (13.5).

13.12.6 Local authorities have the discretion to fix the tax, subject to central government's powers of capping (13.6 and 13.8).

13.12.7 There is a right of appeal to the valuation tribunal against liability, etc., and for judicial review against the fixing of the level of Council Tax, etc. (13.10).

Chapter 14

Council tax valuation

14.1 Synopsis

14.1.1 Half of the Council Tax is payable for the property element, which is based on the dwelling being allocated to one of eight value bands.

14.1.2 The dwellings are listed in valuation lists and the property element of the Council Tax is based on the entry in the valuation lists.

14.1.3 The dwelling-part of composite hereditaments is placed within one of the eight value bands.

14.2 Introduction

14.2.1 Section 3 of the Local Government Finance and Valuation Act 1991 (abolished by Sch. 14 1992 Act) made provision for the valuation of domestic property in England and Wales for the purposes of the Council Tax in advance of the 1992 Act. The valuations were the responsibility of the valuation officers, renamed for the purposes of the Council Tax as 'listing officers'.

14.2.2 Each property is allocated to one of eight value bands and, subject to variations in the personal element of the Council Tax (see Chapter 13), households in properties in the same bands in a billing authority's area receive the same Council Tax bill.

14.3 'Dwelling' defined

14.3.1 Section 3 (2) of the Local Government Finance Act 1992 defines 'dwelling' as any property which:
(a) by virtue of the definition of a hereditament in s. 115 (1) of the General Rate Act 1967, would have been a hereditament for the purposes of that Act if that Act remained in force; and
(b) is not for the time being shown or required to be shown in a local or a central non-domestic rating list in force at that time; and
(c) is not for the time being exempt from local non-domestic rating for the purposes of Part III of the Local Government Finance Act 1988.

14.3.2 According to Sales, 1996 (at 2-334),

This must be the most elaborate and negative legislative definition of dwelling or ... dwelling house ever achieved.

14.3.3 It is necessary to investigate each category in order to achieve full understanding.

14.3.4 The definition of 'hereditament' in s. 115 of the 1967 Act was:

property which is or may become liable to a rate, being a unit of such property which is, or would fall to be, shown as a separate item in the valuation list.

14.3.5 In fact, under the 1967 Act (as today), it was necessary to refer to the case law outlined in 3.3 in order to provide a workable definition. Thus, a hereditament must be:
(a) capable of definition;
(b) a single geographical unit;
(c) capable of separate occupation; and
(d) used for a single purpose.

14.3.6 In the case of a hereditament which is a composite hereditament for the purposes of Part 3 of the Local Government Finance Act 1988, the part of the hereditament which is domestic property is to be domestic property for the purposes of the 1992 Act valuations. Legislation, therefore, ensures that where part of a hereditament is exempt the Uniform Business Rate because it is domestic, that part becomes liable to the Council Tax, by definition.

14.3.7 Section 3 (4) of the 1992 Act excluded the following from the definition of domestic property, except where they form part of a larger property which is itself domestic property under the definition:

(a) a yard, garden, outhouse or other appurtenance belonging to or enjoyed with property used wholly for the purposes of living accommodation; or

(b) a private garage which either has a floor area of not more than 25 square metres or is used wholly or mainly for the accommodation of a private motor vehicle; or

(c) private storage premises used wholly or mainly for the storage of articles of domestic use.

14.3.8 The Secretary of State may by order amend, or substitute another definition for, any definition of 'dwelling' property which is effective in England and Wales for the purposes of the valuation (s. 3 (6) 1992 Act) and may prescribe that anything which would otherwise be one dwelling be treated as two or more dwellings and vice versa (s. 3 (5) 1992 Act).

14.3.9 The Council Tax (Chargeable Dwellings) Order 1992 (SI 1992 No. 549) gives the listing officer discretion to treat a property which is occupied as more than one unit of separate accommodation as one dwelling. In exercising the discretion, the listing officer must have regard to all the circumstances of the case including the extent, if any, to which the parts of the property separately occupied have been structurally altered.

14.3.10 This legislation allows for a house with a so-called granny flat to be treated as a single dwelling for the purposes of the Council Tax (see *Batty v. Burfoot* (1995)).

14.3.11 Caravan pitches and boat moorings are chargeable dwellings, while certain categories of domestic property may be prescribed as exempt (see 14.7).

14.4 Basis of valuation

14.4.1 Paragraph 6 (1) of the Council Tax (Situation and Valuation of Dwellings) Regulations 1992 (SI 1992 No. 550) defines the 'value of any dwelling' as follows:

the value of any dwelling shall be taken to be the amount which, on the assumptions mentioned ... below, the dwelling might reasonably have been expected to realise if it had been sold in the open market by a willing vendor on the 1st April 1991.

14.4.2 The assumptions are:

(a)   that the sale was with vacant possession;

(b)   that the interest sold was the freehold, or in the case of a flat, a lease for 99 years at a nominal rent;

(c)   that the dwelling was sold free from any rent charge or other encumbrance;

(d)   that the size, layout and character of the dwelling, and the physical state of the locality, were the same as at the date the valuation was made;

(e)   that the dwelling was in a state of reasonable repair;

(f)   in the case of a dwelling the owner or occupier of which is entitled to use common parts, that those parts were in a like state of repair and that the purchaser would be liable to contribute towards the cost of keeping them in such a state;

(g)   in the case of a dwelling which contains fixtures to which this sub-paragraph applies, that the fixtures were not included in the dwelling. Such fixtures are those:

(i) which are designed to make the dwelling suitable for use by a physically disabled person, and

(ii) which add to the value of the dwelling;

(h)   that the use of the dwelling would be permanently restricted to use as a private dwelling; and

(i)   that the dwelling had no development value other than the value attributed to the permitted development.

14.4.3   The definition of value cited in 14.4.1 is similar to that used as the basis for compensation under s. 5 of the Land Compensation Act 1961 (which provides for compensation for land taken following compulsory acquisition to be based on: 'the amount which the land if sold in the open market by a willing seller might be expected to realise').

14.4.4   From compensation and other cases where similar phraseology has been used, the definition of the value of a dwelling for the purposes of Council Tax can be further explained.

14.4.5   'In the open market' has been held (*IRC v. Clay and Buchanan* (1914) at p. 888) to mean that the property is 'offered under conditions enabling every person desirous of purchasing to come in and make an offer'.

14.4.6   In particular, the words 'willing seller' have been the subject of interpretation in case law. Thus, 'a "willing seller" is "one who is a free agent", not "a person willing to sell his property without reserve for any price he can obtain for it" ' (ibid. cited by Davies 1984 p. 137). A 'willing seller' is 'assumed to be willing to sell at

the best price which he can reasonably get in the open market'
(*Trocette Property Co. v. Greater London Council* (1974)).

14.4.7   'A price which the land is "expected to realise" ' implies a reference to
'the expectations of properly qualified persons who have taken pains
to inform themselves of all the particulars ascertainable about the
property, and its capabilities, the demand for it and the likely buyers'
– in short, the professional opinion of competent valuers. (*IRC v. Clay
and Buchanan* (1914), cited by Davies, 1984, p. 137)

14.4.8   *Alterations of the List*
In the case of a valuation carried out for the purpose of an
alteration of the valuation list resulting from a material reduction
in the value of the dwelling, the following assumptions (reg. 6 (3)
Council Tax (Situation and Valuation of Dwellings) Regulations
1992 (SI 1992 No. 550) are to be made:
(a)   that the physical state of the locality of the dwelling was the
same as on the date from which the alteration of the list would
have effect; and
(b)   that the size, layout and character of the dwelling were the
same:
   (i) in the case of an alteration resulting from a change to the
   physical condition of the dwelling, as on the date from which the
   alteration of the list would have effect; or
   (ii) in a case where there has been a previous alteration of the
   valuation list, as on the date from which that alteration had effect;
   or
   (iii) in a case where in relation to the dwelling, there has been a
   relevant transaction, not resulting in an alteration of the valuation list, as on the date of that transaction; or
   (iv) in a case to which more than one of sub-paras (i) to (iii)
   above apply, as on whichever is the latest of the dates there mentioned; and
   (v) in any other case, as on 1 April 1993.

14.4.9   For the purposes of (b) (iii) above, a 'relevant transaction' is a sale
of the fee simple, a grant of a lease for a term of seven years or
more or a transfer on sale of such a lease (s. 24 1992 Act).

14.4.10   *Composite hereditament*
In the case of a composite hereditament, the value of the dwelling
is taken to be that portion of the relevant amount which can
reasonably be attributed to domestic use of the dwelling (reg. 7 (1)
ibid.). 'Relevant amount' means the amount which the composite
hereditament might reasonably be expected to realise on the
assumptions mentioned above (14.4.2).

14.5   Valuation list

14.5.1   There is one valuation list for each billing area, and other arrangements for the list should be similar to those used for the Uniform Business Rate. The Commissioners of Inland Revenue appointed a listing officer (i.e. the valuation officer renamed for the purposes of the Council Tax) for each billing authority who is responsible for the valuation list for that authority.

14.5.2   Section 22 (1) 1992 Act requires a listing officer for a billing authority to compile and maintain a valuation list for that billing authority's area. The contents of the valuation lists are to comply with the Council Tax (Contents of Valuation Lists) Regulations 1992 (SI 1992 No. 553) (s. 23 1992 Act).

14.5.3   The list was to be compiled on 1 April 1993 and to come into effect on that date (s. 22 (2) 1992 Act). It is to be maintained so long as necessary (ibid. (9)). There is no provision for updating entries to the list.

14.5.4   A copy of the draft valuation list was sent to each billing authority before 1 December 1992 and the billing authorities were required to publicise the contents of the list, including personal notification of the band in which each dwelling appeared (s. 22 (6) (7) 1992 Act).

14.5.5   There is a right (s. 28 ibid.) to information about the 'state of a list' from both the listing officer and the billing authority, and there is a right to make transcripts or to have photocopies taken at a reasonable charge.

14.5.6   Under s. 23 of the 1992 Act, the valuation list shows (for each day for which it is in force) each dwelling in the billing authority's area, its reference number and the appropriate valuation band (see 14.6) (Council Tax (Contents of Valuation Lists) Regulations 1992 (SI 1992 No. 553). In addition, the list must identify any composite hereditaments liable to the Uniform Business Rate, part of which consists of domestic property.

14.5.7   Where a list has been altered, the period for which or the date from which the alteration takes effect must be shown and also whether the alteration was made in compliance with a decision of a valuation tribunal or the High Court.

14.5.8   An omission from the list of any matter required to be included will not invalidate the list (s. 23 (4) 1992 Act).

14.5.9   A dwelling which spans the boundary of two or more billing authorities is listed in the area of the billing authority in which the greater part of the building (not the land) is situated (reg. 3 (1) Council Tax (Situation and Valuation of Dwellings) Regulations (SI 1992 No. 550)).

14.5.10   Further details about valuation lists can be found in *The Council Tax: Practice Note No. 1 – Valuation Lists* (revised August 1993) (prepared by the Department of the Environment, the Welsh Office, and associations of local authorities).

14.6   Banding

14.6.1   All dwellings which are liable to Council Tax are placed in one of eight valuation bands. The bands which currently apply s. 5 (2) 1992 Act are as shown in Table 14.1:

Table 14.1 Council Tax bands

Valuation band	Range of values
England	
A	Not exceeding £40,000
B	Exceeding £40,000 but not exceeding £52,000
C	Exceeding £52,000 but not exceeding £68,000
D	Exceeding £68,000 but not exceeding £88,000
E	Exceeding £88,000 but not exceeding £120,000
F	Exceeding £120,000 but not exceeding £160,000
G	Exceeding £160,000 but not exceeding £320,000
H	Exceeding £320,000
Wales	
A	Not exceeding £30,000
B	Exceeding £30,000 but not exceeding £39,000
C	Exceeding £39,000 but not exceeding £51,000
D	Exceeding £51,000 but not exceeding £66,000
E	Exceeding £66,000 but not exceeding £90,000
F	Exceeding £90,000 but not exceeding £120,000
G	Exceeding £120,000 but not exceeding £240,000
H	Exceeding £240,000

14.6.2   The Secretary of State has the power to vary the range of values within the bands and to substitute other valuation bands for those currently in force (s. 5 (4) 1992 Act).

14.6.3   Average property value
The value bands were constructed around the average property values in the country (i.e. England or Wales) in which the properties are located. Thus, banding was carried out separately for properties in England and Wales, by reference to the average property value for the country in question.

14.6.4   Composite properties
Composite properties, which are partly domestic and liable to Council Tax and partly non-domestic and liable to rates, have their domestic part allocated to a band. Council tax paid will depend on that banding and the number of adults resident (see 14.4.10).

14.6.5   Relative burden of taxing
The different bands are defined so that the government's intended relativity of taxing can be established, i.e., a taxable couple living in a dwelling in the highest band pays twice as much as a couple living in a property in the middle band; a couple living in a dwelling in the lowest band pays two-thirds of the bill for a middle-band property. Overall, the bill for couples in the highest band is three times as much as that for couples in the lowest band in the same local authority area (s. 5 (1) 1992 Act). The bands in between represent steps between the lowest and the highest, with the proportions being as shown in Table 14.2:

Table 14.2 Relativity of Council Tax liability

Band A	Band B	Band C	Band D	Band E	Band F	Band G	Band H
6	7	8	9	11	13	15	18

The numbers represent the relative proportions of the Council Tax bill which are paid by taxpayers whose properties fall within the different bands (s. 5 (1) 1992 Act).

Thus, taxpayers who live in Band A properties pay two-thirds of the amount paid by taxpayers who live in Band D properties and taxpayers who live in Band H properties pay twice the amount paid by taxpayers who live in Band D properties.

14.6.6   Review of bands
The Local Government Finance and Valuation Bill proposed that bands should be regularly reviewed to take account of changes in house prices, and the Department of the Environment is proposing a reserve power to order an area revaluation in cases where there has been significant differential movement in the values of different sorts of properties.

14.6.7   Of course, neither of these will reflect the variations in house prices between different regions of the country nor will a review of bands reflect the variations in house prices between different kinds of dwellings (semi-detached, detached, terraced, etc., houses, flats and maisonettes – purpose-built or converted, caravans, etc.).

14.6.8   However, there is no indication at present (April 1996) that there is any political will to introduce a revaluation of dwellings for the purposes of the Council Tax.

14.6.9   *Disabled occupiers*
Under the Council Tax, any alterations made to a property for a disabled person which increase the value of that property are disregarded in assessing the band into which the property should fall (reg. 6 (1) (g) and (4) Council Tax (Situation and Valuation of Dwellings) Regulations 1992 (SI 1992 No. 550)).

14.6.10   *New dwellings*
New properties are assigned to bands on the basis of comparative values at a specified date and the experience of the valuer of the banding of similar properties.

14.6.11   Where a valuation list is altered, the valuation list must show the date from which the alteration has effect and, where appropriate, that the alteration was made in compliance with the order of a valuation tribunal or the High Court (Council Tax (Alteration of Lists and Appeals) Regulations 1993 (SI 1993 No. 290)).

14.6.12   *Structural improvements*
Because the tax is based on broad bands, improvements to a property do not necessarily lead to a change in its original band.

14.6.13   *Second homes*
The consultation document (HMSO, 1991, p. 15) recognised that all properties impose costs on local authorities whether they are continuously occupied or not. Properties such as second homes and houses occasionally let require the protective and maintenance services provided by local authorities, and this is the justification for subjecting such properties to the Council Tax.

14.6.14   Where a property is not anyone's main residence, the Council Tax bill is reduced by the maximum of two personal discounts (see 13.4). The owner of a 'second home', therefore, pays the full property element of the tax, but enjoys the full personal discount, and thus only half the normal Council Tax bill is payable.

14.7   Exempt dwellings

14.7.1   The Council Tax (Exempt Dwellings) Order 1992 (SI 1992 No. 558) (as amended) prescribes classes of dwellings for which no Council Tax is payable, as follows:

Class A: vacant dwellings which require or are undergoing structural or other major works to render them habitable, or alterations of a structural nature; or which have been vacant for less than six months from the day on which such works were substantially completed;

Class B: dwellings owned by a charity and which have been unoccupied for less than 6 months;

Class C: dwellings which have been vacant for less than six months;

Class D: unoccupied dwellings which are the sole or main residence of the owner or tenants who are living or being detained elsewhere, such as in prison or mental care;

Class E: dwellings which are unoccupied and are the homes of owners or tenants who are in hospital, a nursing home or residential care;

Class F: unoccupied dwellings following the death of the occupier where either probate has not been granted or less than six months has elapsed since its grant;

Class G: unoccupied dwellings where occupation is prohibited;

Class H: unoccupied dwellings which are being kept for occupation by ministers of religion;

Class I: unoccupied dwellings which where previously the homes of owners or tenants (not within Class F) who have their main residence elsewhere in order to be cared for by reason of their old age, disablement, illness, past or present alcohol or drug dependence, past or present mental disorder;

Class J: unoccupied dwellings which are the homes of people resident elsewhere to take care of others;

Class K: unoccupied dwellings which are the homes of students resident elsewhere for the purposes of their studies;

Class L: unoccupied dwellings which are in the possession of a mortgagee;

Class M: an occupied dwelling if it is a hall of residence for students or run by a charity;

Class N: an occupied dwelling if it is wholly occupied by students;

Class O: an occupied dwelling if it is part of armed-forces accommodation;

Class P: where at least one person occupying a dwelling is a member of the visiting armed forces;

Class Q: dwellings in the hands of a trustee in bankruptcy;

Class R: a dwelling consisting of a pitch or mooring not occupied by a caravan or boat;

Class S: a dwelling occupied only by a person or persons under the age of 21;

Class T: an unoccupied dwelling which forms part of a single property which includes another dwelling which cannot be separately let, e.g. a granny flat in a house;

Class U: a dwelling occupied only by a person (or persons) who is (are) severely mentally impared.

14.8   Billing authorities

14.8.1   A valuation list has been prepared for each billing authority by a listing officer who is responsible for the valuation list for that area.

14.8.2   The listing officer provides copies of, and notification of changes in, the list to the billing authorities.

14.8.3   Billing authorities send out bills; further reference to the administrative role of billing authorities should be made to *Council Tax: Practice Note No. 5 – Administration (including billing and collection)* (revised April 1994) (prepared by the Department of the Environment, the Welsh Office, and associations of local authorities) and *Council Tax: Practice Note No. 9 – Recovery and enforcement* (revised September 1993) (prepared by the Department of the Environment, the Welsh Office, and associations of local authorities)

14.9   Appeals

14.9.1   Appeals against liability (see 13.3) and against the allocation of a dwelling to its band are to the valuation tribunal (Council Tax (Alteration of Lists and Appeals) Regulations 1993 (SI 1993 No. 290)).

14.9.2   No alteration can be made to the valuation list unless:
(a)   since the band was first shown as applicable to the dwelling:
(i) there has been a material increase in the value of the dwelling and a relevant transaction (see 14.4.4) has subsequently taken place;
(ii) there has been a material reduction in the value of the dwelling (unless caused by demolition work which is part of other operations in progress);
(iii) the dwelling has ceased to be a composite hereditament; or

(iv) in the case of a dwelling which is a composite hereditament, there is an increase or decrease in its domestic use;

(b)   a different valuation band should originally have been determined as applicable to the dwelling or that the band shown in the list was not that originally determined; or

(c)   an order of the valuation tribunal or High Court requires an alteration.

(Reg. 4 (1) and (2) Council Tax (Alteration of Lists and Appeals) Regulations 1993 (SI 1993 No. 290), as amended by the Council Tax (Alteration of Lists and Appeals) (Amendment) Regulations 1994 (SI 1994 No. 1746).)

14.9.3   Under reg. 5 (ibid.) a billing authority or interested person may make a proposal to the listing officer if they believe that:

(a)   the list shows a dwelling which ought not to be shown;

(b)   the list fails to show a dwelling which ought to be shown;

(c)   the band attributed to the dwelling by the listing officer is incorrect;

(d)   since the valuation band was first shown as applicable in the list, any of the events described above (14.9.2(a)) has occurred; or

(e)   account has not been taken of a decision of a valuation tribunal or the High Court, regarding a particular dwelling or class of dwelling.

14.9.4   For the purposes of the Council Tax, an 'interested person' is defined (reg. 2 (1) Council Tax (Alteration of Lists and Appeals) Regulations 1993 (SI 1993 No. 290)) as the owner or any person who is substituted for the owner under s. 8 (3) of the 1992 Act (which allows for owners to be identified as the taxpayers in prescribed classes of dwelling such as houses in multiple occupation (see 13.3.11)); any person who would be liable to pay council tax in respect of an exempt dwelling were that dwelling not exempt; or any other person who is a taxpayer in respect of that dwelling.

14.9.5   Where a valuation list is altered, it must show the period for which or, as the case may be, the date from which the alteration has effect and, where appropriate, that the alteration was made in compliance with the order of a valuation tribunal or the High Court.

14.9.6   The procedure is considered in detail in *Council Tax: Practice Note No. 6 – Appeals* (prepared by the Department of the Environment, the Welsh Office, and associations of local authorities).

14.10 Check-list

14.10.1 Council Tax valuations are undertaken by listing officers (14.2.1).

14.10.2 'Dwelling' is defined as a hereditament not liable to the UBR and includes composite hereditaments (14.3).

14.10.3 The value of the dwelling is the amount which the dwelling might reasonably have been expected to realise if sold in the open market by a willing vendor on 1 April 1991, subject to specified assumptions (14.4).

14.10.4 The list can only be altered for specified reasons (14.4.8).

14.10.5 Each dwelling is allocated one of eight bands and it is on its band that the level of the property element of the Council Tax is based (14.6).

14.10.6 Certain dwellings are exempt, e.g. vacant dwellings undergoing structural repair to render them habitable (14.7).

14.10.7 There are limited rights of appeal against the entry of a dwelling in the valuation list (14.9).

Chapter 15

Criticisms of the Council Tax

15.1 Synopsis

15.1.1 The Council Tax is a tax levied by local (billing) authorities on the value of domestic property, paid by the occupier (or, where there is no occupier, the owner), raising money for local authority expenditure.

15.1.2 Like all taxes, the Council Tax should conform to recognised principles of equity.

15.1.3 The Council Tax should be investigated to ensure that it is a fair, up-to-date and socially acceptable system and that it is implemented equitably.

15.2 Generally

As with the UBR, criticisms can be made against the Council Tax at various levels, including the fact that it is a tax on a necessity (see 12.3). This Chapter concentrates on some of the main criticisms of the principles of the tax and its implementation. Reference should be made to texts which deal with the principles of taxing land (and buildings) and, once again, reference should be made to the bibliography included for further reading.

15.3 Principles of the tax

15.3.1 *Personal discount*
The Council Tax combines the principles of the Community Charge (poll tax) to the extent that a discount (of 25% of the overall bill) is given if there is only one (taxable) adult resident in the dwelling (see 13.4).

15.3.2 This can be criticised as being totally at odds with the principle of adjusting tax to the ability to pay. The personal element, which was included apparently for political motives, reflects the wishes of the Conservative government not to abandon totally the principles of the Community Charge (poll tax) when the Council Tax was originally introduced as its replacement in 1993 (see Appendix A).

15.3.3 However, giving a 25% discount where only one taxable adult is resident is also seen as encouraging the under-utilisation of residential property within the taxation system. This is impossible to justify in a country with a chronic housing shortage.

15.3.4 *Quality of valuations*
The method of valuing the dwellings for the implementation of the Council Tax came in for considerable criticism at the time the valuations were undertaken. The valuation of some 23 million domestic hereditaments in England and Wales was carried out between November 1991 and May 1992. This is a phenomenal work load to be undertaken, with accuracy, during a period of only seven months.

15.3.5 Under s. 3 (1) of the Local Government Finance and Valuation Act 1991 (now repealed), the Commissioners of Inland Revenue, who were charged with the duty to carry out the valuations in England and Wales, were able to appoint others who were not in the service of the Crown to assist them in carrying out the valuation exercise, so long as the rules of confidentiality applicable to the Commissioners of Inland Revenue were not breached. Private sector valuers who could show that they had the professional competence, capacity and financial viability to undertake it, were offered the opportunity to tender for two-thirds of the valuation work, which amounted to approximately 14 million valuations.

15.3.6 Following the competitive tendering exercise, 512 contracts were awarded to 337 private firms, divided between 1,052 geographical lots. The Valuation Office Agency undertook the banding of the remaining 9 million properties. The Valuation Office was described (HMSO, 1991, p. 12), as having 'both the experience and the independence to ensure the integrity of the banding process'.

15.3.7 It was announced in December 1991, that the total value of the contract placed was £19.1 million and that, within the private sector, this meant an average price of £1.58 per domestic property valued for the banding of the Council Tax.

15.3.8 The Valuation Office Agency was responsible for maintaining overall quality control and, during the course of this monitoring, the VOA made spot checks on the valuations provided by private practice, although it was reported that in only about five cases where private practice firms made inaccurate valuations were their contracts terminated.

15.3.9 The low cost and speed with which the valuations for the Council Tax were produced brought inevitable questions regarding their accuracy. Despite the assurances of VOA quality control, there have, however, been some well-publicised cases of errors. The government justified the low price paid for the valuations on the grounds that they were not 'valuations' at all, merely the placing of every dwelling into one of eight bands, a process not as exact and therefore not as time-consuming nor as expensive as true 'valuations'. Nevertheless, after the fiasco of the Community Charge (poll tax) (see Appendix A), anything which brings its replacement into disrepute should be strenuously avoided.

15.3.10 *Banding*
Council Tax ignores ability to pay, for example, by placing all properties over £320,000 in England and £240,000 in Wales in the same band. If it is true that those with more disposable income live in more valuable dwellings, then the very wealthiest in the country are not required to pay proportionately more as their disposable wealth (as reflected in the value of their dwellings) increases.

15.3.11 Some dwelling-houses are worth millions of pounds, yet their occupiers and/or owners pay the same level of Council Tax as if they were worth £320,000 (or £240,000 in Wales).

15.3.12 Such taxpayers are, in any event, only required to pay twice the amount paid by more than the so-called average Band D property in the same local authority area (see 14.6.5 and Table 14.2), which is most unlikely to reflect the relative levels of wealth of such occupiers and/or owners. While such control of the relative burden of taxation may be justifiable at the lower end of the property bands, it cannot be justified on the grounds of ability to pay at the higher end of the bands.

15.3.13   Indeed, if such a Band H property is occupied by only one taxable person, the bill is reduced by 25%.

15.3.14   There is no justification for such protection of the most wealthy residents in England and Wales, whether on the grounds of the equity of taxation or on the grounds of social equity.

15.3.15   *Revaluation*
It is not the intention of the government (January 1997) to provide regular and frequent revaluations of dwellings. Banding is recognised (HMSO, 1991, p. 10 and p. 12) as having a number of key advantages:
(a)   the banding system is capable of ensuring that excessive burdens do not fall on a minority of properties;
(b)   undue variations in bills arising from regional variations in property values could be avoided;
(c)   banding should reduce the administrative task involved in the introduction of the tax and allow it to be achieved more quickly;
(d)   it would reduce the likelihood of disputes and appeals;
(e)   there would be no need for regular and frequent revaluations.

15.3.16   Indeed, at present, it is not the intention of central government to provide any revaluations at all. New properties are assigned to bands of the basis on their comparative values at a specified date. The only likely amendment is that central government may decide to vary the values of the bands and, as at March 1996, the Under-Secretary of State for the Environment confirmed that there were no plans to review the bands.

15.3.17   It is well recognised following the public concern demonstrated (following, for example, the seventeen-year life of the 1973 valuation list) that a failure to maintain an up-to-date valuation base will rapidly cause relative unfairness to those taxpayers affected and, without a revaluation, the Council Tax will fall into disrepute. A review of the bands would not allow for relative shifts in value between different kinds of dwellings in different localities to be reflected within the bands. Regular revaluations (ideally at intervals to match those currently undertaken for non-domestic property) is essential to maintain the credibility of the Council Tax.

15.3.18 *Rights of Appeal*
There is, currently, no right of appeal against the inclusion of the dwelling in the wrong band, i.e., against the valuation, unless some additional event takes place.

15.3.19 There was a right to challenge the original banding of the property, but such appeals could only be made between 1 April 1993 and 30 November 1993 (Council Tax (Alteration of the Lists and Appeals) Regulations 1993 (SI 1993 No. 290), reg. 5), and only new owners whose dwelling was not the subject of an earlier appeal are now entitled to challenge the banding under this provision (ibid.). There is, therefore, no right to challenge the banding of dwellings for the majority of the taxpayers.

15.3.20 Rights of appeal by taxpayers relating to valuation exist (Local Government Finance Act 1992, s. 24) only following a material increase or decrease in the value of the dwelling. (All other rights under s. 24 are available only to the listing (valuation) officer, who is required to rectify errors, both valuation and clerical, and to ensure that physical changes in the property and its environment which are reflected in the value of the property are matters which can be taken into account when ensuring the accuracy of the list.)

15.3.21 However, all Council Tax valuations must be correct as at 1 April 1993, so any subsequent changes in value due to economic factors must be ignored. Physical changes which affect value can be taken into account, and the property placed into the band appropriate to the value of the property in its current state as at 1 April 1993.

15.3.22 With no prospect of revaluation of properties, this will merely exacerbate the unfairness between certain taxpayers.

15.4   Implementation of the tax

15.4.1 Capping
One of the principles underlying the introduction of the Council Tax was the principle that no one should pay excessive levels of local authority taxes. Central government implements this principle by imposing a limit on the level of local government spending, which effectively limits the level of Council Tax which can be raised.

15.4.2   While it can be argued that such restraint is important, it runs totally counter to the fundamental principle underlying the Council Tax, that of local democratic accountability. If local democratic accountability is to be truly observed, then local government should be accountable to the local electorate via the ballot box and have no national controls imposed on the level of its spending or tax-raising powers.

15.4.3   Respect should also be given to local government by central government, as this would allow local government to conduct its affairs in harmony with central government and not, as at present, within the financial strait-jacket imposed from Westminster. The UK should, therefore, respect the fundamental principle of local democratic accountability by permitting local government to raise adequate financial resources.

15.4.4   Failure to do so contributes to the impossible position into which local government is currently forced by central government – a position which gives all the semblance of promoting local democratic responsibility but none of the substance.

15.5   Conclusions

15.5.1   These criticisms are fundamental, in that they relate to the fairness of the tax liability both between Council Tax payers and between Council Tax payers and other taxpayers.

15.5.2   With the slump in residential house prices during the early and mid-1990s, there is little public concern at present about the issue of appeals against the valuations and a future revaluation. Once the residential selling market becomes more buoyant, and taxpayers perceive other properties, within the same tax band as their own, selling for dramatically more money than their own properties are worth, public opinion will demand a change in the fundamental principles of the tax.

15.5.3   It is such a pity that the lessons of the pre-1990 rating system have not been learned by the legislature and that the local government system of taxation (and, by implication, local government itself), will inevitably be brought into disrepute again, with the public being forced to press for a change which could so easily be made now.

Appendix A – from rates to Council Tax (1990–1993)

Background to reform

Before the rates reforms which took effect on 1 April 1990 and which introduced the Uniform Business Rate (UBR) (described in Chapters 1 to 12), local authorities levied a property tax called the general rate on all non-domestic and domestic hereditaments in their area. Indeed, at this time, no distinction was made between the two.

The rating system had been under scrutiny for some years and, because of the problems in assessing domestic property to rates which were evident in the 1973 revaluation (i.e. the paucity of open-market rental evidence for domestic property), reform of the rating system became urgent. However, the decision to abolish domestic rates was a political one, and had been part of the Conservative party's policy since the 1970s.

There was widespread political and public criticism of the general rate (see HMSO, 1976, for example) and the Conservative government which took office in 1979 was committed to abolishing rates. However, it was necessary that abolition be accompanied by a replacement system of raising revenue for local government expenditure, and this proved a major political problem for the Conservative government in the early 1980s.

Several government and non-government committees considered the question of reforming the rating system and the alternatives which could replace it (e.g. HMSO, 1981 and HMSO, 1983), and all of them supported the retention of a reformed rating system, possibly supported by another source of local authority revenue. It was the Conservative government's commitment to abolish rating which made any solution based on a reformed land tax or rating system politically unacceptable.

In 1986, the Conservative government produced *Paying for Local Government* (HMSO, 1986), a White Paper which proposed a poll-tax-based replacement for the rates levied on residential property.

A Local Tax

The Community Charge, which was introduced with effect from 1 April 1990, was a kind of poll tax proposed by the Conservative government in its White Paper *Paying for Local Government*, which was published in January 1986.

The White Paper set out certain criteria which were considered fundamental to any source of local authority funding. In order to achieve those criteria, a variant of the poll tax, called the Community Charge, was outlined.

(Since a poll tax is, by definition, a tax on every individual, and since under the Community Charge not all individuals in the UK were liable to pay the tax, it was argued that the Community Charge should not be called a poll tax, but a variant of a poll tax – although such a distinction is fine, and popular opinion favoured the use of 'poll tax' rather than the official title.)

Criteria for a local tax

In its 1996 White Paper, the government outlined the changes made during the twentieth century which had made the domestic rating system unsuitable for supporting local authority finance. The changes were:

(a)   a move towards the 'redistributive principle' under which rates are levied generally to pay for services. This can be distinguished from the 'beneficial principle' under which rates were levied on particular groups to share out the cost of providing an amenity from which they all benefited;

(b)   the transition to an industrial society and the growth of urban development, which created the need for more communal services and new forms of local government. The process of consolidating many separate 'rates' resulted in the setting, by a local authority, of a single general rate based on the value of property;

(c)   the changes in the nature and responsibilities of local authorities and in the relationship between local authority responsibilities and the role of central-government funding through the Exchequer grants;

(d)   the expansion of the franchise, i.e., the right to vote in local elections, which, today, is available to any resident Commonwealth citizen or citizen of the Irish Republic, over the age of 18.

In the light of these important changes in the franchise and in the scope of local authority services, the White Paper looked again at the suitability of domestic rates as a local tax. It identified three main tests which a local tax should satisfy:

(a)   technical adequacy;
(b)   fairness; and
(c)   local democratic accountability.

a.   Technical adequacy
In the 1981 Green Paper *Alternatives to Domestic Rates*, the technical adequacy of local taxes was considered. The 1981 Green Paper argued, in particular, for the importance of cost-effective administration, for compatibility with sensible and tight budgeting, i.e. proper financial control, and for the need for any local tax to fit in with the overall national tax system.

The yield should be predictable, not lumpy, permit the local authorities to plan ahead and to make 'small discrete changes in the tax rates' (HMSO, 1986, pp. 19–20). In addition, the increase in tax should be 'perceptible' (i.e. the taxpayer should be required to pay out of earned income and thereby be aware of the level of tax) so as to ensure effective financial control. Finally, any new local tax introduced should be suitable for all tiers of local government.

b.   Fairness
In advocating 'fairness', the 1986 White Paper criticised the failure of rates to take account of both the 'beneficial principle' and the 'redistributive principle'. Also, it pointed out that domestic rates do not properly reflect the use made by a household of local services. In addition, the White Paper remarked on the swing away from property-based services, e.g., water, gas, electricity, fire service and police protection, towards more personal services, e.g., education, social services and libraries, which, it stated (ibid. p. 20), account for over 60% of current local authority expenditure.

A tax based on the value of properties obviously would not be closely or consistently related to the use of such services (ibid. p. 20).

Rates were also criticised in relation to the 'redistributive principle': in setting out the relationship between rate bills and incomes, it was pointed out that 'those with the lowest incomes are liable to much higher rates in relation to their income.' This is, of course, why the rate rebate scheme was introduced into the pre-1990 rating system, so that those with very low incomes paid no rates at all. No definition of 'fairness' was included in the 1996 White Paper.

c. Local democratic accountability
The Government considered that:

whether a tax is conducive to proper local democratic accountability is now of crucial importance. It is the key to an approach to local government finance that rests on responsible local spending decisions and a reduction in central Government intervention. (ibid. p. 21)

Central government, therefore, made accountability its main priority and identified the following features as being capable of securing accountability of local authorities to their electorate:

(a)   the tax base should be wide so that the burden is not unfairly concentrated on too few;
(b)   a substantial proportion of electors should have a direct financial interest in the decisions of their authority;
(c)   there should be a clear link between changes in expenditure and changes in the local tax billing.

Alternatives to domestic rates
The alternatives to the pre-1990 system of rates were debated in the 1981 Green Paper. However, with the redefined objectives of technical adequacy, fairness and local democratic accountability, the 1986 Green Paper re-examined domestic rates, local sales tax, local income tax, and 'the way forward – a community charge'.

a. Reformed rating system
Domestic rates, reformed so that the tax would be based on capital values, and with some minor alterations, were rejected on the grounds that they did not take into account the relatively large number of the electorate who contributed to rates only indirectly (e.g. income tax-paying adults who were not ratepayers) or who paid no rates because of the 100% rate rebate scheme and for whom, therefore, rates were imperceptible. Rates, it was considered, fell on too few shoulders.

Rates were levied on a tax base which was not self-adjusting. Rates were levied on rateable values (defined in substantially the same way as at present), which needed to be periodically up-dated. Failure to update the rateable values regularly and frequently led to sudden and large shifts in the tax burden. Even regular frequent revaluations would mean that rates were increasing for reasons other than local expenditure. In addition, rateable values varied throughout the country, so that there were large differences between local authorities in rateable value per head, which required compensatory arrangements through central government grants.

b.   Local sales tax

A local sales tax is not perceptible by electors and would be complex to administer. The yield would be lumpy and uneven throughout the local authority areas, with regional shopping centres benefiting at the expense of centres with fewer retail outlets. There is the danger of cross-border shopping and no evidence to suggest that people shop within their local authority area anyway. Government grants would continue to be necessary to compensate for the differences, with problems of apportioning the yield between authorities.

c.   Local income tax

A local income tax would represent the final shift away from the 'beneficial principle' and would require a separately administered rebate scheme, as well as an equalising grant to even out the yield between different authorities. In addition, a local income tax was contrary to central government's policy to reduce the burden of income tax; it would not underpin local accountability; and it was not considered to be an appropriate tax to be administered by local authorities, since local authorities should have no role in controlling the economy. It is not perceptible and does not support local accountability, since only those wage-earners who vote would be paying for local government expenditure.

The substantial changes in the tax administration to allow local authorities to set, collect and administer a local tax would be complex, expensive for employers and breach the confidentiality of the tax return.

d.   Poll tax

The Community Charge, as a variant of the poll tax, was accepted in the 1986 White Paper as being better than rates with regard to local accountability, since:

[rates] are no better related to ability to pay than a flat-rate charge would be ... and they are less well related to use of local authority services which now more closely reflects the number of people in a household than the value of the property occupied. Furthermore they fall on a narrow section of the local voting population. (ibid. p. 24)

The government, therefore, decided to introduce a new flat-rate charge for local services, payable at the same level by all the adults resident in a local authority area. Such a charge, it was considered, would be more perceptible than rates, and the level, determined by each local authority, would be the local authorities' only controllable source of income.

Moving from rates to a community charge marked a major

change in the direction of local government finance towards the notion of charging for local authority services.

The 1986 White Paper considered that the Community Charge provided a closer reflection of the benefit from modern people-based services than a property-based tax.

As before, the bulk of local authority services continued to be funded from national taxes, including the Uniform Business Rate, so that those on higher incomes continued to pay proportionally more for local authority services via the Exchequer grants.

Criticisms of a poll tax

The l981 Green Paper (HMSO 1981) concluded that a flat-rate charge, payable by all adults, would meet all the technical criteria, would produce sufficient yield, would be suitable for all tiers of government and would be conducive to proper financial control. In rejecting a poll tax as a replacement for rates, it was stated (as reported in the 1986 Green Paper p. 26) that:

... the tax would be hard to enforce. If the electoral register were used as the basis for liability, it could be seen as a tax on the right to vote. A new register would therefore be needed, but this would make the tax expensive to run and complicated, particularly if it incorporated a rebate scheme.

However, by 1986, 'These problems are not insuperable' (HMSO, 1986, p. 26).

The Community Charge

The Community Charge was introduced by the Local Government Act 1988 to take effect on 1 April 1990 alongside the Uniform Business Rate. The Community Charge was a daily flat-rate charge, levied on every adult resident in the local authority's area. Each such charge-payer was identified in a Community Charge Register which was compiled and held by the local or charging authority.

Fixing the level of the Community Charge

The Community Charge was fixed, levied, collected and spent by local, called charging, authorities. Each authority was able to fix the Community Charge at a level which would provide sufficient funds to make up the deficit of revenue received from the Exchequer through grants and receipt of its share of the Uniform Business Rate.

However, central government maintained its power to limit the ability of the charging authorities to fix the Community Charge by 'capping' the amount of Community Charge.

There were, in fact, three Community Charges: Personal Community Charge, Standard Community Charge and Collective Community Charge.

Personal Community Charge

Every adult over the age of 18 was liable to the Community Charge levied by that authority for each day the individual's sole or main residence was within the charging authority's area. The following were exempt from the Community Charge: persons in detention (unless imprisoned for failing to pay the Community Charge); visiting forces; severely mentally-impaired individuals; children over the age of 18 for whom child benefit is payable; members of closed religious communities; patients whose sole or main residence is a hospital, residential care home or nursing home; certain care workers and those of no fixed abode.

Students in full-time education, who were considered to be resident at their place of education, paid only 20% of the charge.

Also liable to pay at least 20% of the charge were those on social security benefits. This was a substantial shift away from the previous 100% rate relief for residential occupiers on low income. Although the Personal Community Charge was a tax on the individual, the amount payable was dependant on the location of that individual's sole or main residence.

Standard Community Charge

The Standard Community Charge was designed to ensure that those owners of houses which were not anyone's sole or main residence contributed to the cost of local authority services. The Standard Community Charge was levied on freeholders or leaseholders of domestic property in respect of which no one paid a Personal Community Charge. In this way, owners of second homes (and builders who were attempting to sell newly-completed houses) contributed to the income of local authorities which provided services for the benefit of those properties.

The Standard Community Charge was set at a proportion of the Personal Community Charge, which was multiplied by either 0; 0.5; 1; 1.5 or 2, depending on the reason why the dwelling was not occupied. Certain classes of dwellings were identified as attracting a multiplier of 0 to the Personal Community Charge (and therefore no Community Charge at all was payable). Such dwellings included dwellings which are undergoing structural repair. Charging authorities were free to levy up to twice the Personal Community Charge on owners of so-called second homes, if they chose.

Collective Community Charge

The Collective Community Charge was paid by landlords of

houses in multiple occupation where it was considered difficult to enforce the Personal Community Charge on the occupiers. The occupiers of such properties would have paid rent to the landlord and the landlord was required to collect not only rent but the Personal Community Charge as well. The landlord would then be liable for the amount of the Personal Community Charge for each occupier of the dwelling.

This was introduced as an administrative measure for ensuring payment from those properties where the occupiers were likely to change address at frequent intervals. It mirrored pre-1990 arrangements made for the collection and payment of rates by an owner of such property and was generally not considered to be unreasonable.

Replacement of the Community Charge

The public outcry, hostility and political repercussions over the introduction of the Community Charge, resulted in November 1990 in Michael Heseltine, the new Secretary of State for the Environment, being given the task of providing a politically-acceptable replacement for the Community Charge six months later, in November 1990.

Criticisms of the Community Charge

The Community Charge marked a significant shift from a property-based tax to a people-based tax; from the original concept of rates as a means of raising money from the wealthy to support the poor to the 'everyone pays something' philosophy characteristic of Margaret Thatcher's period as prime minister; and from the (relative) freedom of local authorities to impose a level of tax appropriate for its local needs to strict central government controls over the level of local tax raised and thereby local authority spending.

There can be no doubt that the Community Charge failed as a tax. Criticisms of any tax depend on the criteria which are established as a base or which is generally acceptable to the tax-paying public. In the case of the Community Charge, the 1986 White Paper established as criteria, technical adequacy, fairness (as defined above) and local accountability. It is patently obvious that the Community Charge failed its own criterion of technical adequacy. 'Fairness' was never appropriately defined within the White Paper and local accountability was never a serious goal for a local authority tax which was limited by central government's capping powers.

It is generally accepted (Page, 1976) that the social acceptability of a tax is vital from a taxpayer's point of view. In Britain of the 1990s, it was very obvious from the day the Community Charge was introduced that requiring everyone to pay the same amount was intrinsically unfair.

When a poll tax was last introduced in England in 1381, the result was the Peasants' Revolt, a variation in the rate of tax between rich and poor and a loss in revenue (Rose, 1988, p. 81). The scenes reported on by the media in early 1990, which were generally representative of popular opinion, proved that the Community Charge was not socially acceptable.

There was no pretence at adjusting the tax bill to the ability to pay within the Community Charge. Indeed, there was an absolute requirement that all charge-payers must contribute at least 20% of the charge, regardless of income, or lack of it. It is a generally-held principle of taxation that those in different financial circumstances pay different amounts of tax (vertical equity). It made no moral sense to require a millionaire to contribute the same level of tax as the ubiquitous elderly widow living alone on her state pension. The rating system had at least recognised that the poorest occupiers should have all their rate liability paid for them through the housing benefit, since taking money from those who have none is illogical, expensive and punitive.

The few available exemptions and the relief of up to 80% for those on social security, student grants, etc., were totally inadequate, given that the level of full Community Charge payable in some areas exceeded £500 per annum.

Another obvious problem which the Community Charge posed was that of certainty of yield. People are more mobile than property and the loss in revenue due to mobility and the inevitable failure of some chargeable adults to re-register and pay the charge reduced the yield achieved by the tax, and increased the costs of administration, both of which increase the rate of tax.

In fact, administration of the Community Charge in Scotland in 1989–90 was estimated by legal authorities to have cost £31.8m (including registration work but excluding operating the rebate system), while the cost of collecting rates in 1988–89 was £17.8m.

Similarly, an estimate of the total cost of collecting and administering the Community Charge in 1990–91 is £400m, compared with the estimate for rate collection in 1989–90 of £200m.

The principle of widening the tax base to all electors so that accountability is improved would have been more socially acceptable if an element of ability to pay was reflected in the charge by providing a 100% rebate for the poorest electors and varying the rate of charge according to the charge-payers' ability to pay.

Alternatively, if the average annual charge had been reduced to a nominal level rather than the ultimate £560 in one local authority area, and had been supported by an alternative source of local authority revenue, then the tax would probably have become more socially acceptable, payment would have been more certain and perhaps the Community Charge would have survived.

By 23 April 1991, Michael Heseltine, the then Secretary of State for the Environment, was announcing the proposals for the replacement of the Community Charge, introduced barely twelve months previously, with the Council Tax and, by September 1991, the Valuation Office was producing valuations for the Council Tax.

Council Tax

Following the failure of the Community Charge (or poll tax) (see above), the defeat of Margaret Thatcher in the Conservative Party leadership election in November 1990 and the appointment of John Major as Prime Minister, Michael Heseltine, as the new Secretary of State for the Environment, was given the task of providing a politically-acceptable replacement for the Community Charge.

The consultation paper (HMSO, 1991, p. 2) identifies the following five principles which guided the work of the review to find a replacement for the Community Charge:

Accountability – the new system should ensure that local people see a link between what they are being asked to pay and what their local authority is spending;

Fairness – the new tax should be perceived as fair by the public;

Ease of collection – administrative arrangements for collection and enforcement should be as straightforward as possible. The costs of administering a local tax should be reasonable in relation to its yield;

Equitable distribution of the burden – the principle that most adults should make some contribution has been widely accepted;

Restraint – a system of local taxation should not allow tax bills to become too high, either because of unreasonable levels of spending by local authorities, or because the system imposes a disproportionately high burden on any individual or household compared with others.

It was intended (ibid. pp. 2–3) that domestic taxpayers in Great Britain should continue to meet the same level of local authority expenditure under the Council Tax as under the Community Charge, i.e., about 14%.

On 23 April 1991, Michael Heseltine, the Secretary of State for the Environment, gave details to the House of Commons of the proposed Council Tax which would take effect on 1 April 1993. The enacting statute, the Local Government Finance Act 1992, provides for certain local authorities to levy and collect the Council Tax; to abolish the Community Charge; to make provision with respect to local government finance (including provision with respect to certain grants by local authorities); and for connected purposes.

The tax (which is described in detail in Chapters 13 and 14) can be summarised as follows:

(a) The Council Tax has a property and a personal element. Each household receives a single bill on the assumption that it consists of two people. Households with only one adult are entitled to a 25% personal discount;

(b) The amount of Council Tax payable varies according to the value of the property but only within a limited range. Properties in England and Wales are allocated to one of eight bands. There was no precise valuation of every house or flat, nor is there perceived to be a need for regular general revaluations (as at January 1997);

(c) The grant mechanism remained broadly unchanged;

(d) People on low incomes receive rebates in addition to any discount to which they are entitled. The maximum rebate for those at the income-support level is 100%. Students, student nurses, apprentices and youth-training trainees are automatically entitled to personal discounts;

(e) The amount of Council Tax payable relates to how much the council spends. Where councils spend more, all households pay more. The same percentage increase applies to all household bills. Similarly the benefit of low spending is passed through to all Council Tax payers;

(f) Central government ensures that no one faces an unreasonable increase in their bill between one year and the next by the use of appropriate transitional relief during the early years of the new system.

It is estimated that Council Tax preparations cost £156m up to the date of its introduction (1 April 1993). Central Government agreed to pay £86m in support of estimated revenue costs of £114.6m and to issue supplementary credit approvals in support of capital expenditure of up to £41.2m.

The full implications of the long-term effects of the poll tax are still being realised. The 1991 census figures were abandoned because some 1.2 million people were unaccounted for. One of the official reasons for this is the poll tax!

Appendix B – History of rating

Poor Relief Act 1601
Provision for the levying of a local tax, at regular intervals, over the whole country.
Appointment of Overseers of the Poor to collect, spend and administer taxation.
Provide work for able-bodied unemployed, and give relief to those incapable of working.
Assess and levy a poor rate on every inhabitant and occupier to finance the above (no basis of assessment laid down).
Right of appeal to the Quarter Sessions.

Sir Anthony Earby's Case 1633
Parish poor rate could be levied only on property within that parish.
If rates are levied on an occupier, they cannot also be levied on the landlord of the same property.

Poor Rate Exemption Act 1833
Exempted places of worship of denominations other than Church of England – not rated in practice under the 1601 Act.

Poor Law Amendment Act, 1834
Administrative Unit was re-organised into Unions of Parishes.
Responsibility for poor relief was transferred to the Board of Guardians of the Poor in each Union.

Parochial Assessment Act 1836
First definition of 'net annual value'.

Poor Rate Exemption Act 1840
Abolished the taxation of stock-in-trade.
Abolished the liability of 'every inhabitant' within a parish to liability to rates.

Scientific Societies Act 1843
Scientific Societies, i.e., those instituted for the purposes of

science, literature or the fine arts exclusively, are exempted from rates.

Union Assessment Committee Act 1862
Provided a definition of 'gross estimated rental'.
Provided for Assessment Committees to supervise assessments.
Required Overseers to prepare and keep a valuation list.
Gave the right of objection to assessments.

Union Assessment Committee Amendment Act 1864
The making and hearing of an objection, and the failure to obtain relief were conditions precedent to an appeal to Special or Quarter Sessions.

Valuation (Metropolis) Act 1869
Separate gross value and rateable value definitions.
First scale for repair definitions.
(Until 1963, London had a separate system of assessment and collection of rates from the remainder of England and Wales. The system was progressively reduced until, for the 1963 List, London was almost completely assimilated to that of the rest of the country.)

Rating Act, 1874
Extended rateable property to all woodlands, sporting rights when severed from the occupation of land, and all mines.
Repealed so much of the 1601 Act as related to the taxation of an occupier of saleable woodlands.

Advertising Stations (Rating) Act 1889
Governed the rating of advertising stations.

Agricultural Rates Act 1896
Granted a 50% relief from rates for agricultural land.

Agricultural Rates Act 1923
Granted a 75% relief from rates for agricultural land.

Rating and Valuation Act 1925
New definitions of 'gross value', 'net annual value' and 'rateable value' to promote uniformity.
Abolition of Overseers and a transfer of their functions to rating authorities.
Repeal of Union Assessment Acts.
Requirement for quinquennial revaluations.
Institution of the 'general rate'.

New system of appeals.
Rating of owners, where more convenient than rating occupiers.

Rating and Valuation (Apportionment) Act 1928
Defined industrial and freight-transport hereditaments.

Local Government Act 1929
Functions of Unions of Parishes transferred to County or County Borough Councils.
Complete exemption from rates for agricultural land and buildings.
Granted a 75% relief from rates to industrial and freight-transport hereditaments.

Tithe Act 1936
Extinguished tithe rent charges and reduced the rating of tithes to a few tithes of a special nature.

Local Government Act 1948
Responsibility for the making, keeping and amending of the valuation list transferred to the Inland Revenue.
Introduction of a system of Exchequer Grants.
Introduction of Local Valuation Courts as the courts of the first instance, replacing assessment committees.
Provided for the rating of advertising rights, when let out or reserved to anyone other than the occupier.

Laing (John) & Sons, Ltd. v. Kingswood Assessment Area Assessment Committee (1949)
Recognition of the four essential ingredients of rateable occupation.

Valuation for Rating Act 1953
Introduced for the 1956 valuation lists, until 1963, a 1939 rental-value basis for dwelling-houses.

Rating and Valuation (Miscellaneous Provisions) Act 1955
(Replaced Poor Rate Exemption Act 1833.)
Granted relief from rates for charities, welfare structures and sewers.

Rating and Valuation Act 1957
Granted a one-fifth reduction in the assessments of commercial and business properties, until 1963.

Local Government Act 1958
Reduced the relief from rates for industrial and freight-transport hereditaments to 50%.

Rating and Valuation Act 1961
Abolished relief given to industrial and freight-transport hereditaments.
Introduced a formula method of valuations for premises occupied by statutory undertakers, etc.
Enabled rating authorities to forego rates on hereditaments for a short time.

Diplomatic Privileges Act 1964
Replaced common-law immunity from payment of rates with statutory immunity.

Local Government Act 1966
Introduced discretionary power of the rating authority to collect rates on unoccupied buildings.
Reduced the rate in the pound for dwelling-houses.
Introduced statutory 'tone of the list' for proposals after 2 December 1965.
Introduced rebates for residential occupiers with low incomes.
Repealed so much of the 1601 Act as to leave only 'occupiers' liable for rates.
Abolished the rating of tithes.

General Rate Act 1967
Consolidated all previously current legislation.

General Rate Act 1970
Widened the scope of evidence which could be produced to justify the assessment of dwelling-houses.

Mines and Quarries (Valuation) Order 1971
Granted a 50% reduction in rateable value from 1 April 1972, for mines and quarries, including those of the NCB.

Rating Act 1971
Exempted 'factory farms' for the intensive raising of stock from rating.

Local Government Act 1974
Minor structural alterations to dwelling-houses not assessable until revaluation.
New rebate scheme for occupiers of dwelling-houses.
Amendment to powers of rating authorities to rate unoccupied properties (abolition of the seven-year rule).
Introduction of the penal surcharge on unoccupied commercial buildings, levied on the owner.

Layfield Report on Local Government Finance, May 1976
Recommendations included:
Capital value rating for domestic properties.
Rental value retained for non-domestic properties.
Regular and frequent (three-yearly) revaluations.
Speeding up of appeals to local valuation courts.
Appeal to Lands Tribunal on points of law, precedent and complex cases.

Rating (Caravan Sites) Act 1976
Valuation officer's discretion to rate leisure caravans on site separately.

Rating (Charity Shops) Act 1976
Amends s. 40 General Rate Act 1967, so that charity shops are exempt.

Rating (Disabled Persons) Act 1978
Repealed s. 45 General Rate Act 1967.
Relief is given by rating authority in rate rebates on dwellings adapted for disabled persons and institutions for the disabled.

Valuation Lists (Postponement) Order 1978
Postponed the requirement for new valuation lists for another year (postponed annually until 1980) to allow consideration to be given to altering the present rating system.

Local Government, Planning and Land Act 1980
Abolished requirement for quinquennial revaluations. Future revaluations to be ordered by the Secretary of State, who may specify hereditaments to be revalued and allow adjustment of assessments of hereditaments not revalued.
On next revaluation, all hereditaments except 'domestic hereditaments' will be valued to NAV (net annual value) direct.
On next revaluation, valuations will be made as at a specified time, subject to certain assumptions regarding 'relevant factors'.
Fish farms derated.
Extension of domestic rate relief to hereditaments which include some residential accommodation.
Extension of right to pay rates by instalments to occupiers of non-domestic hereditaments.
Secretary of State may vary proportion of rates withheld during first year of a list pending settlement of a proposal to alter that list.
Rating authority may vary percentage reductions given to owners who are rated in place of the occupier (s. 55 General Rate Act 1967).

Secretary of State may direct that s.17A General Rate Act 1967 (surcharge provisions) shall cease to have effect, and, having done so, may re-introduce those provisions by order.

Secretary of State may vary three (or six) month period of non-occupation before rates paid on unoccupied newly-built premises.

Rating Surcharge (Suspension) Order 1980
Provisions of s. 17A General Rate Act 1967 (introduced by Local Government Act 1974) ceased to have effect as from 1 April 1981. (Note that the provisions are not abolished, only suspended.)

Alternative to Domestic Rates December 1981
Government Green Paper offering 'for public discussion the results of a review by the Government of possible local revenue'.

Local Government Finance Act 1982
Abolished supplementary rates and supplementary precepts.
Required rates and precepts to be made for complete financial years.
To provide for the making of substituted rates and substituted precepts.
Regulations for challenging the validity of the rate and precepts.
Relief from rates for enterprise zones – additional provisions.
New auditing provisions.

Social Security and Housing Benefits Act 1982
Abolishes previous rate rebate scheme (under General Rate Act 1967 and Local Government Act 1974).
New scheme grants rate rebates along with rent rebates and rent allowances as Housing Benefits.

Rates – Proposals for rate limitation and reform of the rating system, August 1983
Government White Paper proposing reform of the rating system and 'rate capping' of individual local authorities together with a general power to limit rate increases for all authorities.

Rating (Exemption of Unoccupied Industrial Hereditaments) Regulations 1984
Amends Sch. 1 General Rate Act 1967 in that from 1 April 1984 local authorities' powers to levy rates on empty industrial property is suspended.

Rates Act 1984
Introduced 'rate capping', i.e., selective and general limitation of rates and precepts.

Duty on rating authorities to consult industrial and commercial rate payers in fixing rate.

Duty to inform ratepayers of proposed expenditure and any increase/decrease in rates.

Exempts non-domestic hereditaments not in active use.

Discretion of valuation officer to rate moorings separately.

Rating (Exemption of Unoccupied Industrial Storage Hereditaments) Regulations 1985

Amends Sch. 1 General Rate Act 1967 in that from 1 April 1985 local authorities' powers to levy rates on empty warehouse property is suspended.

Paying for Local Government, January 1986

Government White Paper proposing:

abolition of domestic rates, to be replaced by a 'community charge' poll tax;

revaluation of all non-domestic hereditaments to rateable values as at 1 April 1988, the lists to take effect from 1 April, 1990;

simplification of grant system to local authorities.

Local Government Act 1988

Abolished domestic rating with effect from 1 April 1990.

Introduced the statutory provisions for the Community Charge which came into effect on 1 April 1990, as a flat-rate charge on all adults levied by charging authorities.

Repealed the General Rate Act 1967 while retaining many of its provisions.

Laid down framework for Uniform Business Rate and its implementation the details of which were introduced by subsequent statutory instruments.

Local Government and Housing Act 1989

Amended the Local Government Finance Act 1988.

Uniform Business Rate introduced, 1 April 1990

Uniform Business Rate (UBR) introduced.

Revaluation (as at 1 April 1988) took effect.

Community Charge introduced to replace domestic rates.

A New Tax for Local Government, April 1991

Government discussion document proposing the Council Tax to replace the Community Charge (poll tax).

Local Government Finance Act 1992

Implemented the Council Tax to replace Community Charge from

1 April 1993 as the source of local authority revenue from domestic occupiers.

Wood Report, 1994
Wood Report on the rating of plant and machinery published, followed by Valuation for Rating (Plant and Machinery) Regulations 1994 (SI 1994 No. 2680), which amend the existing regulations, with effect from 1 April 1995.

1 April 1995
1995 revaluation takes effect, with a valuation date of 1 April 1993.
Phasing arrangements introduced for the following five years.
UBR introduced for Scotland.

Local Government and Rating Act 1997
Provided mandatory or discretionary relief from non-domestic rates for general stores etc., in rural settlements.
Abolished rating of sporting rights.
Abolished exemption from rates for Crown occupation and ownership.
Exempted hereditaments occupied by a visiting force and its headquarters.

Appendix C – The Bayliss Report

The Royal Institution of Chartered Surveyors, 1996. *Improving the System* (The National Committee on Rating)

In 1996, the Royal Institution of Chartered Surveyors (RICS) published *Improving the System*, which is the report of the National Committee on Rating, chaired by Jeremy Bayliss, and which included representatives from the Incorporated Society of Valuers and Auctioneers, the Institute of Revenues, Rating and Valuation, the Rating Surveyors Association and representatives of various other interested groups including the British Chambers of Commerce and the Confederation of British Industry.
The Committee took as its terms of reference:

To review the business rating system throughout England, Scotland and Wales, in particular: to examine ways in which the administration, efficiency, effectiveness and equity of the rating system might be improved; and to make recommendations.

The Report made the following recommendations (which are reproduced here with the kind permission of the RICS):

Comprehension, harmonisation and efficiency of the rating system

1. Rating legislation for England and Wales, for Scotland and for Northern Ireland should be consolidated (para. 1.3.2).
2. Similarly, the administrative procedures for England and Wales, Scotland and Northern Ireland should be harmonised as far as possible, taking the best from each country (para. 1.4.4).
3. Billing authorities should at revaluations supply each ratepayer with a notice which has been prepared jointly with valuation officers, giving the provisional new rateable value and an estimate of the likely rate bill (para. 1.3.7).

4.   All personnel working within the rating system should be properly trained in the use of IT (para. 1.5.9).

5.   The Government should require that all future IT projects within the rating system should meet minimum standards of compatibility and be interactive (para. 1.5.11).

6.   The provision of a central source for valuation tribunal decisions to be made readily available to the public be investigated (para. 1.5.14).

7.   The VOA [Valuation Office Agency],valuation tribunals and billing authorities must maintain adequate resources and structures to fulfil their statutory duties properly (para. 1.5.15).

The rate liability

8.   There should be a single UBR for the whole of Great Britain which should remain subject to the RPI annual change ceiling (para. 2.2.1).

9.   The UBR should not automatically follow the change in the rate of inflation but, as is already provided for in legislation, that change should set a ceiling figure and not be treated as mandatory (para. 2.2.5).

10.   Consultations between local government and local business would be made more meaningful if they took place as part of the process by which budgets are set (para. 2.3.7).

11.   Both the upward and downward arrangements for transition should be eliminated by the year 2000 (if not earlier) with no further 'transition upon transition' (para. 2.4.9).

12.   Billing authorities should be given as an incentive to further efficiency and motivation and to ensure the maintenance of a broad tax base, the right to retain some part of the rate income which they collect (para. 2.5.2).

13.   Rate income collected and retained by billing authorities should be excluded from the per capita redistributive process (para. 2.5.3).

14.   There is no case for further Special Authorities. Any proposal for the extension of such status should be based on research (para. 2.6.3).

Valuation and revaluations

15.   From 2000 onwards revaluations should take place at three-yearly intervals (para. 3.3.6).

16.   The antecedent period in the long term should be one year and in the short term 18 months (para. 3.4.8).

17.   Self-assessment for the rating of non-domestic property is not a practical proposition (para. 3.5.4).

18.   It is not appropriate at the present time to amend the definition of 'rateable value' (para. 3.6.4).

19.   It would not be appropriate or practical to change to a capital value basis for non-domestic rates (para. 3.7.1).

20.   Further investigation is necessary to identify whether there are good reasons for introducing a banding system for rateable value (para. 3.8.7).

21.   Formula-assessed hereditaments should be returned to a conventional basis of valuation by the year 2000 (para. 3.9.2).

22.   There should be a central rating list for Scotland for consistency throughout Great Britain (para. 3.9.4).

23.   A standing committee should be established to provide regular reviews of the rating of plant and machinery (para. 3.11.1).

Appeals

24.   The time limit for the transmission of proposals from valuation officers to valuation tribunals in England and Wales should be reduced to one month (para. 4.3.7).

25.   A detailed study of a 'civil magistracy' to incorporate a lands valuation division as an alternative appellate body to valuation tribunals should be undertaken (para. 4.4.3(iii)).

26.   There should be a right of appeal direct to the Lands Tribunal as an appeal and not only as an arbitration (para. 4.4.6).

27.   A consultation process should be instituted to formulate long-term listing programmes by valuation tribunals (para. 4.5.7).

28.   The procedures for selecting members of valuation tribunals should be changed (para. 4.6.3).

29.   There should be appropriate training and refresher programmes for valuation tribunal members (para. 4.6.5).

30.   There should be appropriate training in advocacy and the preparation of proofs of evidence for valuation officers and ratepayers' representatives who present cases and give evidence before valuation tribunals (para. 4.6.6).

31.   An agreed code of procedure in respect of valuation tribunal appeals in England and Wales should be promoted (para. 4.7.1).

32.   The proposed changes to the Lands Tribunal Rules relating to administrative procedures should be implemented as soon as possible (para. 4.8.9).

33.   The new Lands Tribunal Rules should require that the

documentation to be lodged with the pleadings should include a statement of agreed facts (para. 4.8.10).

34. The complement of the Inland Revenue Solicitor's Office should be supplemented without delay or private practice solicitors or Counsel instructed by the Inland Revenue (para. 4.8.11).

35. The proposed changes in respect of the VOA's [Valuation Office Agency's] position as regards the award of costs in rating appeals should be introduced as soon as possible (para. 4.8.15).

36. The Lord Chancellor's department is urged to monitor rating appeals to the Lands Tribunal to ensure that they are not delayed because of a lack of resources (para. 4.8.16).

37. A simplified procedure for rating appeals to the Lands Tribunal should be introduced (para. 4.8.20).

38. There should be a restriction on the retrospective effect of proposals by introducing a six-months 'effective rate liability date' (para. 4.9.4).

39. Action should be taken against unprofessional 'rating advisers' throughout the UK through education of the public and, if need be, by restricting the right of audience at valuation tribunals (para. 4.10.3).

Collection and refunds

40. All rate demand notices should have the basic information of the amounts demanded in a standard format (para. 5.2.1).

41. Legislation should be enacted to provide enforceable liability for rate payment on property occupied by insolvent businesses continuing to trade (para. 5.3.4).

42. Rates demanded retrospectively after a lengthy delay which is not the fault of the ratepayer should be payable by instalments (para. 5.4.2).

43. The valuation officer should be required to notify the ratepayer on the same day that the rating list is altered (para. 5.5.4).

44. A code of good practice for communication between billing authorities and valuation officers should be promoted by the Government (para. 5.5.5).

45. Refunds on overpayment of rates should be made to ratepayers within eight weeks of appeal settlements or determinations (para. 5.5.6).

46. Failure to make refunds within the eight weeks should carry a penalty of substantially higher interest payments to ratepayers (para. 5.5.7).

47. A billing authority should have the right to recover from the

VOA [Valuation Office Agency] such higher interest in the event that a failure by a valuation officer prevents it from complying with the time limit for the payment of refunds (para. 5.5.7).

48. Rate refund notices should give full details of the reasons for the refund and computations of the amount in a standard format (para. 5.5.9).

49. The alteration to the Non-Domestic Rating (Payment of Interest) Regulations already proposed by the DoE [Department of the Environment] to make interest payable on all refunds should be laid before Parliament without delay (para. 5.5.13).

50. Rights regarding payment of refunds should be clearly explained to ratepayers and a detailed statement of interest payable should be sent at the same time (para. 5.5.14).

51. The repayment of overpaid rates, together with interest, should become automatic (para. 5.5.15).

52. Guidance notes on all aspects of the rating system to enable ratepayers to understand it should be issued by the Government (para. 5.6.2(i)).

53. Billing authorities should publish customer care policies for ratepayers and the Valuation Tribunal Service should publish a 'Charter' (para. 5.6.2(ii)).

54. The VOA [Valuation Office Agency] in undertaking to clear large numbers of bulk classes of property must not prejudice the speedy clearance either of the more complex cases, or of those of high value (para. 5.6.3).

55. The need for meetings between the valuation officers and billing authorities will become increasingly more important with the rationalisation of the VOA [Valuation Office Agency] local office network and these should take place at regular intervals (para. 5.6.5).

Exemptions and reliefs

56. The Government should set up a Committee to review the whole question of exemption from the non-domestic rates (para. 6.2.5).

57. Relief on the grounds of hardship, where this is in the interests of the local community, should remain at the discretion of the billing authorities (para. 6.3.3).

58. Billing authorities should emphasise on rate demand notices the relief available to ratepayers on the grounds of hardship (para. 6.3.3).

59. The need for a definition of 'hardship' should be considered (para. 6.3.7).

60.   Billing authorities should be reminded that it is illegal to resolve not to use discretionary powers (para. 6.4.3).

61.   Billing authorities should have the power to grant discretionary rate relief retrospectively (para. 6.4.6).

62.   Ministers should consider the extent of mandatory relief given to charity shops (para. 6.5.4).

63.   Special rate relief for rural shops should be approached with caution as more effective direct assistance can be afforded through proper application of the hardship relief provisions (para. 6.6.3).

64.   If special rate relief is granted to rural shopkeepers it should be funded centrally and not from the rate pool or by Council Tax payers (para. 6.6.6).

65.   Consideration should be given to the proposal by the Federation of Small Businesses for banding relief for small businesses (para. 6.7.5).

66.   Consideration should be given, as an alternative to Recommendation 65, to a lower UBR for all 'small properties' within the current transitional thresholds and the abolition of a separate level of transition for such properties (para. 6.7.6).

67.   The Non-Domestic Rating (Unoccupied Properties) Regulations 1989 should be updated in line with current planning law (para. 6.8.3).

68.   The initial grace period for unoccupied property rate should be extended to six months in line with the Council Tax provisions and the charge thereafter reduced to 25% of the full liability (para. 6.8.4).

69.   An initial grace period should apply to the unoccupied part of apportionments under s. 44A and should extend to six months followed by a charge of 25% of the full liability, where the occupied part is not otherwise exempt (para. 6.9.3).

70.   A review of the operation of s. 44A should be undertaken and consideration given to making the relief mandatory instead of discretionary (para. 6.9.7).

Appendix D – Valuation date

History

Following the decision in *Barratt v. Gravesend Assessment Committee* (1941), in which it was held that a person making a proposal to alter the rating list had to show that he was aggrieved when he made that proposal, it was necessary to determine what rent a hypothetical tenant would pay for the hereditament on the date of the proposal for a yearly letting from that date. Thus, once a rating list comes into force, the valuation date is the date of the proposal. However, in the earlier case of *Ladies Hosiery & Underwear Ltd. v. West Middlesex Assessment Committee* (1932), the company established that its assessment was too high in comparison with the assessments of other comparable properties in the locality, but had to admit that its assessment was not in excess of the rent that might reasonably be expected for it on the statutory terms at the date of the proposal. The Court of Appeal refused to reduce the assessment to the level of the comparable properties, deciding that correctness of assessment should not be sacrificed to uniformity.

The only remedy available to the company was to make proposals on the comparable hereditaments to ensure that their assessments were increased to the same level as that of the appeal hereditament!

Thus, unfairness was to be remedied by seeking uniform correctness and not by establishing uniform incorrectness.

The rules established by these two cases (that the valuation date is the date of the proposal and that correctness must not be sacrificed to uniformity), would have caused few problems had there been frequent and regular revaluations and had there been no periods of high inflation. But with infrequent revaluations (1939, 1956, 1963, 1973, 1990, 1995) and with periods of rapid and high inflation (particularly during the 1960s–1980s), the relativities established at the beginning of the rating lists quickly became out of date and ratepayers sought ways of putting pressure on the system to allow for the appeals procedure to remedy the unfairness produced. This problem did not exist when a new list came into effect, because all hereditaments are valued as at the same date. The problem only arises during the life of a list and is aggravated by infrequent revaluations and inflation.

For example, if shops in a high street were valued at £1 per m² in terms of zone A as at 1 April 1939, and after the war shop rentals increased tenfold, then if one of the shops was enlarged in such a way that it had to be revalued, the value to be applied to the new area would be £10 per m², at a time when all the other shops were valued at only £1 per m².

The occupier of the one altered property could not argue that £10 per m² was excessive because it would be the current rental value as at the date the shop was enlarged. Unless all the other shops had their assessments increased to that level, unfairness was unavoidable and constant increases in rateable values were never ending.

It was this large-scale use of proposals to solve the problem that the Lands Tribunal attempted to halt by calling the then current rental value, 'a rent paid in extreme scarcity by persons in dire necessity' (*Thomas v. Cross* (1951) and *Moore v. Rees* (1952)). By giving increased attention to the evidence of comparable assessments, the courts ensured that the level of value established by the original entries in the list, or 'tone of the list' would be maintained during the life of the rating list.

A statutory tone of the list was originally established by the Local Government Act 1966, under which every hereditament was valued in its state as at the date of the proposal but at values not exceeding those prevailing when the list came into effect.

However, during the life of the 1973 list, another problem was identified. The then relevant statutory provisions (s. 20 General Rate Act 1967) required every hereditament to be valued to the tone of the list on the assumption that the hereditament was in the same 'state' as at the time of the proposal and any relevant factors (the mode or category of use; the volume of trade of a public house etc.) including value were those subsisting at the valuation date (1 April 1973); and the locality in which the hereditament is situated was in the same 'state', so far as concerns the other premises situated in that locality and the occupation and age of those premises, the transport services and other facilities available and other matters affecting the amenities of the locality. The 1973 list lasted 17 years, and during that time inflation and other factors affected the relative values of different property types. Occupiers used s. 20 to establish an opportunity to reduce their rating assessments by successfully arguing that 'state' did not mean physical state, but any factor affecting the value of the occupation of their premises.

Thus, 'state' was defined to include the legal state (*Sheerness Steel Co. plc v. Maudling (VO)* (1986)), and the changing economic and social state (*Thorn EMI Cinemas Ltd v. Harrison (VO)* (1986)).

However, the major case on the definition of 'state' was *Clement (VO) v. Addis* (1988), which hinged on the effect of the Swansea Enterprise Zone on the assessments of properties bordering on the Zone. The House of Lords said that 'state' was to be given its widest possible meaning so as to include intangible as well as physical factors which affected value. Thus, the House of Lords held that the existence of the Enterprise Zone was a factor which could be reflected in the rating assessment as at the date of proposal.

However, in response to that decision, the then Secretary of State introduced legislation to redefine 'state' to reflect only the physical state of the property and the locality and it is this definition of 'state' which appears in current legislation (Sch. 6 (2)(3) 1988 Act).

Current legislation

Under Sch. 6 (2)(3) of the 1988 Act, the date of valuation for the 1995 rating list is as follows:

(a)   1 April 1993 is the date by reference to which the level of values and 'non-mentioned matters' are to be taken into account;
(b)   1 April 1995 is the date by reference to which 'mentioned matters' are to be taken into account, when valuing for the preparation of the list;
(c)   the date of the proposal or the date the list is altered is the date by reference to which 'mentioned matters' are to be taken into account, when valuing for an alteration to the list.

'Mentioned matters' are:

(a)   matters affecting the physical state or physical enjoyment of the hereditament;
(b)   the mode or category of occupation of the hereditament;
(c)   matters affecting the physical state of the locality in which the hereditament is situated or which, though not affecting the physical state of the locality, are nonetheless physically manifest there; and
(d)   the use or occupation of other premises situated in the locality of the hereditament.

Since the valuation date of 1 April 1993 occurs before the rating lists came into force (on 1 April 1995), this is an antecedent valuation date (or AVD).

For the 1990 rating list, the Secretary of State specified 1 April 1988 as the valuation date, i.e., all rateable values in the rating lists are as at 1 April 1988, with 1 April 1990 as the date on which the list came into force, and for the lists which will take effect on 1 April 2000, the antecedent valuation date is 1 April 1998.

All valuations, whether made for the purposes of compiling the lists or for altering the lists, must be correct as at values which apply at the antecedent valuation date (currently, 1 April 1993).

However, such matters as the physical state of the hereditament are to be taken as either at the date the lists are compiled when valuing for the purposes of compiling the 1995 revaluation (i.e. 1 April 1995), or at the date the list is altered by the valuation office or at the date of the proposal if the list is subsequently altered.

Thus, while using 1 April 1993 values, a hereditament is valued for the purposes of the revaluation in its physical state as at 1 April 1995 or in its physical state at the date of a proposal or the date when the list is in fact altered.

For example, shop properties in the high street are worth £300 per m² Zone A under full repairing and insuring terms as at 1 April 1993. On 1 April 1995, Shop A has an area of 100 m² in terms of Zone A. In the absence of any other factors which affect its value, the rateable value must be: £300 × 100 m² = £30,000.

However, if on 1 April 1997, the shop is enlarged to an area of 150 m² in terms of Zone A, then the rateable value must be altered to: £300 × 150 m² = £45,000.

Thus, while the size is changed to the actual size at the date of the proposal, the unit value of £300 remains the same as that applicable in 1 April 1993.

In this way, the level of value or tone of the list remains the same during the life of the list, but the 'mentioned matters' as at the date of the proposal vary the rateable value. (See Appendix E for more details of how shops are zoned.)

In 1992, the VO modified the view that only physical factors can affect rateable values by accepting that the increase in vacant office space since 1988 amounted to an extraordinary material change in circumstance. The increase in vacant office space was due in part to changing economic conditions, but also to the increase in new space becoming available – a physical change in circumstances. In such circumstances, reductions in rateable value, to reflect in part the reduced rents being achieved, could be made.

Appendix E – Zoning

Introduction

Zoning is a method of analysis applied to shop areas and rents, in order to use the rental evidence of one (or several) shop(s) as a basis for fixing the rental value of another shop. Although often referred to as a method of valuation, zoning is a method of analysis which produces a rental value per unit which can be compared with units of similar properties. It is, thus, a variation of the comparative and rental methods of valuation, producing a unit of comparison based on rental evidence. Zoning assumes that the front part of a shop is the most valuable part, with the value decreasing away from the frontage. Analysis of the ground-floor sales area reflects this principle by a division of the sales area into zones of relative value, as shown:

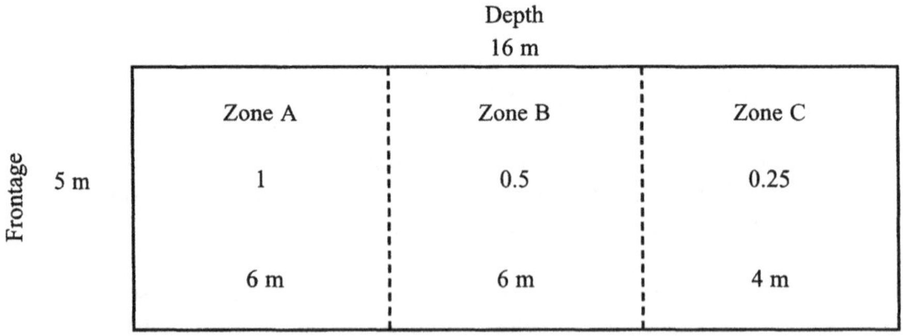

Depth
16 m

Zone A		Zone B		Zone C	
1		0.5		0.25	
6 m		6 m		4 m	

Frontage 5 m

Figure E1

Zones A and B each have a depth of 6 metres, with Zone C comprising the remainder of the ground-floor sales area.

The relative values are achieved by 'halving back', i.e., whatever the value attributable to Zone A (the unit price), the value of

Zone B will be one-half of that unit price; and the value attributable to Zone C will be one-half of the value of Zone B.

These relative values (1:0.5:0.25) are applied to the areas of the three zones, so that the area of ground-floor sales can be analysed into an area in terms of Zone A.

If this analysis is carried out for a shop which is rented, then a rental value in terms of Zone A can be calculated.

This rental value in terms of Zone A (the unit price) can then be used as a unit of comparison for other similar shops which have been analysed on the same basis.

Example:

Shop X: Frontage   5 metres
           Depth   16 metres

Tenure: let last month on a three-year lease on full repairing and insuring terms for £20,000 p.a. (the stated terms).

Shop Y: Frontage   6 metres
           Depth   24 metres

Tenure: Owner-occupied.
Shops X and Y are comparable in all other material aspects.

Analysis/zoning of area of shop X:

Zone A: 5 m × 6 m at 1   30
Zone B: 5 m × 6 m at 0.5   15
Zone C: 5 m × 4 m at 0.25   5
Area in terms of Zone A   50 m²

Rental Analysis:

Shop X has a rental value of £20,000 on the stated terms. It therefore has a unit price of:

$$\frac{20,000}{50} \text{ per m}^2 \text{ Zone A}$$

i.e. £400/m² Zone A

The unit price of £400/m² Zone A can now be used to value comparable shops which have been analysed on the same basis.

Analysis/zoning of area of shop Y:

24 m

6 m	Zone A	Zone B	Zone C
	6 m	6 m	12 m

Zone A: 6 m × 6 m at 1   36
Zone B: 6 m × 6 m at 0.5   18
Zone C: 6 m × 12 m at 0.25   18
Area in terms of Zone A   72 m²

Valuations:

Knowing that Shop X has a unit price of £400/m² in terms of Zone A and that Shops X and Y are comparable, it follows that Shop Y has a unit value of £400/m² in terms of Zone A, and is, therefore, worth:

$$400 \times 72 = £28,800 \text{ per annum}$$

if let on full repairing and insuring terms, on a three-year lease.

Remember:
(a) always compare like with like.
(b) as you devalue, so you value.

Origins of zoning

The zoning method of analysis is thought to have been devised by the late Sir Herbert Trustram Eve in 1917, when 'parlour' shops and retailing patterns were little changed from previous centuries.

Objectives of zoning

The recognised objectives of the zoning method of analysis were cited in *Trevail (VO) v. C. & A. Modes, Ltd.* (at p. 199) as being:

(a) to provide a means whereby all shop-rating assessments could be made on a correct and uniform basis; and
(b) to provide a means whereby the rating assessments of large shops could be reached by an examination of the rents of smaller shops;

and in *J. Sainsbury Ltd. v. Wood (VO)* (at p. 211):

(c) to apply the evidence of rents to shops of varying depths to arrive at a pattern of values to be attributed to other, broadly similar, shops where no rental evidence was available.

Theory

The theory supporting the zoning method is 'that the front parts of shops in a busy high street were more valuable than the rear parts' because of their 'proximity to the bustle of the city pavement and the inducement offered to window shoppers' (ibid., p. 257).

Having enticed customers into the shop, the front sales area is used to attract customers deeper into the shop. Thus, the frontage and the first area of depth (Zone A) is the most valuable, with the rest of the sales area being, normally, valued at a proportion of the value applied to Zone A. The theory was generally true when it was devised and probably remains true today for certain sizes of shops in certain localities with certain occupiers. It will not be true, however, if applied to an out-of-town superstore.

The theory, however, supports market evidence that if a shop of, say, $80m^2$ is worth £25,000 per annum (on certain terms), a shop of twice that size is not necessarily worth twice that rental value (on the same terms), giving a logical justification for the apparent discrepancy.

Zoning practice

Zoning practice is not standardised throughout the country, with tradition dictating varying lengths for the zones; one, two or three zones plus a remainder; and, in rare cases, there is no traditional use of a zoning method of any kind, e.g., Brighton (*B.H.S. Ltd v. C.B. of Brighton & Burton (VO)* (1958).

Except in the use of 'arithmetical' or arbitrary zoning, it is generally accepted that the correct zoning depth in any given locality is based on analysis of different zone depths. The depth which gives the most consistent results will be correct for shop units in that locality.

It is, therefore, a question of 'trial and error', using reliable, comparable rental evidence in any given case, and, of course, this means that each locality must be treated separately and the results can not be compared with other localities.

This raises problems where rental evidence is not available and reliance must be placed either on theory (i.e., arithmetical zoning), or the use of an apparently comparable locality and the unit price derived from analysis carried out on rents within that locality.

Variations in the number of zones include:

1.   Zone A 6.1 metres (20 feet)
     Zone B 6.1 metres (20 feet)
     Zone C remainder (20 feet)

   (*W.H. Smith & Son Ltd v. Clee (VO)* (1977))

2.   Zone A 4.57 metres (15 feet)
     Zone B 7.62 metres (25 feet)
     Zone C remainder

   (*F.W. Woolworth & Co. Ltd v. Moore (VO)* (1978); *Janes (Gowns) Ltd. v. Harlow Development Corporation* (1980))

3.   Zone A 18.29 metres (60 foot)
     Remainder

(*Trevail (VO) v. C & A Modes Ltd* (1967))

As indicated above and stated in *Trevail (VO) v. C. & A. Modes, Ltd.* (p. 199), 'There is no hard and fast rule'. But once a decision has been reached on the number and depth of the zones, the same zones must be adopted for all comparables, whether they are to be analysed or valued. As you devalue, so you value.

Regardless of the number or depth of zones used, the 'halving back' of zones of ground floor sales areas seems to be universal, i.e., if Zone A has a value of £100/m², then Zone B has a value of £50/m², and Zone C a value of £25/m².

It is, however, recognised that the 'halving back' of the remainder may not always be appropriate, and it is rare for ancillaries to be halved back.

Ancillaries are often separately valued, or given a value which is a fraction of the Zone A figure, e.g., one-sixth, or one-tenth.

Ancillaries

These are any additional accommodation for which rent or other consideration has been or could be paid but which is not ground-floor sales area.

Ancillaries will include stock rooms, upper floors, basements and even car-parking facilities. According to *Trevail (VO) v. C. & A. Modes, Ltd.* (1967), at p.200, a value must be put on all ancillary accommodation and this value deducted from the rent to provide a balance, which is the value of the entire ground-floor selling space.

Such a value for ancillaries is, presumably, derived from market rents for such accommodation when let separately from ground-floor sales areas.

Deriving rent for ancillaries in this way may not always be appropriate, if it can be shown that such accommodation, when let separately, will not command the same rent as when let with ground-floor sales areas. Again, it should be a question of analysis of market transactions in order to establish the correct approach.

However, it is often the case that the valuer will use a pre-determined relationship between the Zone A value and ancillaries, with the latter being taken at one-sixth or one-tenth of the Zone A rent. Market evidence should always be produced to support whatever treatment is adopted for ancillaries.

Arithmetical or arbitrary zoning

Partly because of its use for rating purposes, the division of ground-floor sales areas into arithmetical or arbitrary zones is common.

This type of zoning ignores the actual depths of the shop properties and the relative values which rental evidence reveals as existing between one zone and another (see Natural zoning, below) and, for simplicity, has been illustrated in the Introduction above.

Before carrying out any analysis, the valuer will know in advance the number of zones to be used, the depth of each zone and the relative value of each zone and any ancillaries.

This knowledge is likely to be based on local practice, and will enable the valuer to compare all shop valuations, rental evidence and unit prices in the locality.

Arithmetical zoning does have the advantage of being easy to apply, and, if used throughout the country, of allowing all shops to be compared on a uniform basis.

It is accepted by the Lands Tribunal as having 'stood the test of time', but, according to Emeny and Wilks, 1984 (p. 262):

Such a method cannot be theoretically supported as a method of valuation, it is nothing more than an arithmetical formula which is bound to produce inconsistent results in most cases.

Incorrect sizes of zones and incorrect relationships between the value of zones will probably produce answers which are quite divorced from reality. Indeed, by arbitrarily altering the depth of the zones and the relationship of the value between zones, rents can be analysed in different ways so that, when applied to a shop for which there is no rental evidence, different values result.

Natural zoning

(See Emeny and Wilks, 1984, pp. 258–261.)

Natural zoning is a method of fixing zone depths and zone numbers by an analysis of rental evidence available on shops in a given centre or parade.

Zone A would be fixed at the depth of the shallowest shop for which a reliable rent is available, Zone B would end at the depth of the next deepest shop for which reliable rents are available, with Zone C ending at depth of the longest shop for which a reliable rent is available. Anything beyond Zone C might usually be treated as being of storage value only.

All rented shops in that shopping centre or parade would be analysed in terms of these zone depths, and the rental values obtained for each zone used to value other shops in the locality.

In any locality, therefore, the depth of the zones is dictated by the depths of the actual shops, and the relationship in value between one zone and another is dictated by the values produced by market evidence.

In this way, natural zoning is a more reliable method of arriving at rental value than arithmetical zoning because it is a unique analysis for each shopping locality, relying both for its values and mode of application on actual rental evidence.

Natural zoning, however, produces rental evidence which is only useful for the locality from which the evidence was gathered. It is not possible to apply the results of the analysis to any other locality.

Tobacco kiosk

The typical tobacco kiosk, located in shopping centres and comprising only a few square metres, is also likely to be valued by the zoning method, in the absence of more reliable rental evidence.

Like the 'large' shops, the tobacco kiosk is not strictly comparable with 'small' shops, so the use of such rental evidence must be treated with the same degree of scepticism and variation as when applied to 'large' shops.

Disabilities

Disabilities are factors which affect the rental value of one shop and which are not reflected in the rental values of otherwise comparable shops, the rents of which are used to value the first shop.

A shop with disabilities will not command as high a rent as shops without disabilities (all other things being equal) and, for this reason, an end allowance must be made to account for the disability.

Disabilities may take the form of changes in floor level, columns, internal walls, or a lack of facilities which could reasonably have been expected and which occur in otherwise comparable properties.

Inconvenient shape is dealt with later.

Quantity allowance

A quantity allowance becomes important when units of rental value derived from 'small' shops are applied to the valuation of 'large' shops, because the two shop types, as defined by size, are not truly comparable.

'Small' shops and 'large' shops form two distinct property types and markets, because the type of trader who requires the retail area provided by a 'small' shop is not going to be interested in renting a 'large' shop, unless he is allowed to forgo paying rent for the excess shop area which he does not require. In contrast, a trader needing a large retail area will not be interested in a 'small' shop.

In the past, the recognition of the lack of demand for space was given in an end allowance, called a quantity or quantum allowance.

Although such an allowance may be given today, it generally reflects the fact that rental values derived from 'small' shops will not produce a true rental value when applied to 'large' shops, without modification. In such circumstances, a quantity allowance may be given to ensure that the rental value fixed is, indeed, the rental value for the actual property being valued, but only if market evidence supports such an allowance (*U.D.S. Tailoring Ltd v. B.L. Holdings Ltd* (1981); *Janes (Gowns) Ltd v. Harlow Development Corporation* (1980); *Trevail (VO) v. C & A Modes Ltd* (1967).

Applied in such circumstances, the amount of a quantity allowance is extremely arbitrary, because it implies that there was other evidence on which to determine rental value, which was ignored in the actual valuation.

A quantity allowance may also be given or confused with an allowance for 'shape'.

In *W. H. Smith & Son Ltd. v. Clee (VO)* (1977), the Tribunal distinguished and defined an 'allowance for shape' (at p. 247) as:

an end-allowance relating to 'size along the frontage', which … is incapable of reflection in the values adopted for the sales area

and distinguished it (ibid) from an 'allowance for size', which is:

an end-allowance related to an overall size in all dimensions, to the extent (if any) that this factor may not have been already adequately allowed for in the values adopted for the various areas of floor space within the hereditament.

If two properties of, say 20 m by 50 m exist in the same shopping parade, with one having a frontage of 20 m but the other having a frontage of 50 m, then in applying the zoning method, the two would have differing rental values, despite the fact that each has the same overall area.

In recognition of such distortions which zoning creates, an allowance for 'shape' may be given, although only to the extent that the frontage:depth ratio does not conform with the normally expected ratio.

In *W. H. Smith & Son Ltd. v. Clee (VO)* (1977), an allowance for 'shape' was given to a hereditament with a frontage:depth ratio of 1:034, compared with other comparable units of between 1:1.78 and 1:4.25. Such an allowance is only given if, and to the extent that, the market would not reflect the full value of increased frontage space, as calculated using the zoning method (ibid.).

Return frontage

A return frontage is a second window or display fronting another, normally secondary, thoroughfare. Return frontages generally exist for corner shops. The theory supporting the zoning method means that the main frontage and the first area of depth (Zone A) is the most valuable because of its proximity to passing customers.

It follows, therefore, that a shop with a return frontage to another pedestrian thoroughfare has an increased value over a shop without a return frontage.

The amount of additional value attached to a return frontage can be calculated from the increased rent paid for a shop with a return frontage as compared with a rent paid for an identical shop unit without a return frontage.

Where market evidence is not quite so convenient, and especially where arithmetical zoning is used, an increase of 10% of the otherwise normal shop rent is usually added. The percentage should be varied depending on the relative retailing importance of the 'side' street, as evidenced by the pedestrian flow.

In extreme cases, especially where 'large' shops have several points of access, confusion may exist as to which frontage is the more valuable. In *Watts (VO) v. Royal Arsenal Co-operative Society Ltd* (1984), the Tribunal accepted that zoning should be taken from the longest, and therefore, the more valuable frontage. It has been known for zoning to be carried out from both frontages, but this raises an even greater doubt over the rental values produced by such a variation in the absence of supporting market rental evidence (Emeny and Wilks, 1984, p. 267).

In fact, this is creating a greater problem in view of the movement away from traditional, rectilinear shopping-centre design to the creation of more interestingly-shaped pedestrian spaces.

Alternative techniques

All valuation techniques are theoretically-based methods of fixing either a rental or capital value for property, in the absence of an open-market transaction.

Under a given set of circumstances, a property has a value to accepted tolerances, which a valuer will seek to establish using recognised valuation techniques.

It is, however, generally assumed that the use of valuation techniques serves to reflect the actions of the open market, and not to lead the open market nor to create an artificial value which would not be achieved in an open-market transaction.

It follows, therefore, that the most logical technique to adopt in fixing a rental value for shop properties is that used by actual landlords and tenants in the open market. It must be queried whether zoning is the technique used by actual landlords and tenants or merely by their valuation-trained advisors.

Turnover of the business

When a landlord and a prospective tenant negotiate the rent to be paid on the grant of a new lease, there are several factors to be taken into account.

Assuming that all other details of the lease are agreed (e.g., term, repair liability, user restrictions, etc.), the amount of the tenant's rental bid will be determined, mainly, by the amount of gross profit which he can make from his occupation of the particular shop unit at that location, and out of which he can afford to pay rent.

'Turnover rents', i.e., rents based on either a percentage of the trading profit, or a combination of base rent plus a percentage of the trading profit, are not uncommon, although they are usually found in shopping centres or concourses owned and managed by a single landlord. The appropriate percentage varies between trades and may be subject to a minimum which equates to a reasonable rent fixed in the more traditional way.

Some retailers use a formula to check rental levels. It relates pedestrian flow past the unit frontage per hour to the length of that frontage, which could be said to be the two factors dictating the tenant's gross profit potential.

There are, of course, considerations of taxation, tenant mix, security of income, competing rental bids, etc., but the rent finally agreed between landlord and tenant will reflect the profitability to the actual tenant of the premises.

It follows, therefore, that if tenants can achieve differing profit levels from different trades, then the most profitable trade which can afford the highest rent should, in theory, always secure the shop premises.

However, particularly in a shopping centre or concourse where a single landlord is keen to secure the success of the centre as a whole, the highest rental bid will not always be acceptable. A good mix of tenants, ensuring the most popular (i.e., profitable) balance of trades is likely to secure the long-term success of the centre or unit, and, for this reason, an analysis of rents paid for modern centres or parades often shows an otherwise illogical pattern.

It is in trying to make sense of such a pattern and in the absence of an actual tenant, the profitability of whose trade can actually be estimated in a given retailing location, that a valuer must resort to valuation techniques, experience and market evidence to fix a rental value.

'Overall price' method

Zoning is based on the hypothesis that the front part of a shop is the most valuable part, with relative values decreasing away from the frontage, on the basis that it is the frontage which attracts passing customers into the shop.

It can be argued, therefore, that zoning should only be applied where shops and trades in fact conform to these criteria.

It is obvious that such criteria will not apply to any shop located away from passing customers and/or where the sight of the front display window will not be the attraction for its customers. In such circumstances, the zoning method is totally inappropriate.

It is generally recognised that zoning is inappropriate for: out-of-town superstores and hypermarkets; supermarkets; 'departmental' stores; and nationally-recognised chain stores. Such places are normally frequented because the customer knows in advance of their existence and of the ranges of goods they sell.

In many cases, it is obvious that frontages are of little or no value to such occupiers, being used either for advertisements, access ways or unadorned walls.

It is also obvious from the layout of the goods within such stores that there is no effort to entice the customer further into the sales area by displaying attractive items close to the frontage. In fact some stores show evidence of displaying frontage attraction in reverse – by ensuring that to get to the attractive goods, the customer must pass the less popular items.

It is generally recognised by valuers that such properties are more realistically valued to a price per square metre overall, the actual overall price being ascertained from an analysis of lettings of comparable properties.

This can be done, either by applying a single overall price to the total net area of the store, or, where there is a constant relationship between sales area and ancillaries, by applying an overall price to the sales area only.

Some valuers consider that ancillaries should be valued on a separate basis and, where they are put to a different use to that applied to the sales area (e.g., offices), should be valued by comparison with similar accommodation elsewhere.

As a method of valuation, the 'overall price' is accepted by the courts (e.g., *J. Sainsbury Ltd. v. Wood (VO)* (1980)), provided that it is an appropriate method for the type of property, and that market evidence can be produced to support both the method and the unit price (*Watts (VO) v. Royal Arsenal Co-operative Society Ltd* (1984)), (*F. W. Woolworth & Co. Ltd. v. Moore (VO))* (1978).

'Frontage' method

An alternative to zoning put forward by Smith, 1984, is a method which, he argues, should produce 'better' answers.

He considers that there is considerable value in the frontage as such, quite separate from the sales area behind it. (See also *B.H.S. Ltd v. C.B. of Brighton & Burton (VO)* (1958) pp. 348–9.)

The 'frontage' method involves calculating:

(a)   the length of the frontage (measured internally) multiplied by a figure consistent with the parade or centre in question; plus
(b)   the total area of the shop multiplied by a different figure consistent with the parade or centre in question.

This method avoids the problem of metrication (see below) if it is applied to rating assessments, and recognises the two main features which most traders require from shop properties, a length of frontage and sales space behind it.

Smith also seeks to show that the 'frontage' method solves many of the problems which zoning creates.

By comparing hypothetical shop properties valued on both the zoning and 'frontage' methods, he concludes that the 'frontage' method deals with the 'tobacco kiosk' problem, the problem of 'shape' where a shop has an excessively large Zone B and remainder as compared with its Zone A, while not significantly altering the valuations of 'regular-shaped shops of reasonable depth'.

Return frontages are dealt with by a simple addition of the return frontage, multiplied by the frontage rate for the street which the return frontage faces.

The method of calculating the multipliers to be applied to the frontage length and shop are calculated using the formula:

$$Fx + Ay = Rent$$

where: F is the length of the frontage;
           x is the multiplier to be applied to the frontage;
           A is the total shop area; and
           y is the multiplier to be applied to the area behind the
           frontage.

Solving the simultaneous equation (using two equations) will give a value for x which, in practice and following a large number of such calculations, would give a pattern of values on which to base a 'frontage' method of valuation for a given locality.

Knowing x (the multiplier to be applied to the frontage), it is possible to calculate y (the multiplier to be applied to the area behind the frontage), by multiplying the frontage by x and adding the area. The total is divided into the rent to produce y.

The theory assumes that the zoning method has produced

reasonable answers for the majority of shops and tests the results of the 'frontage' method against those produced by zoning.

The only real proof for any method is, of course, testing against real rents and, in advocating such a test, Smith proposes that it is in the rating system, particularly with regular revaluations, that such testing should begin.

It is of the greatest importance, however, that any new system should command the respect and consent of the majority of valuers in all branches of the valuation profession.

Zoning for rating purposes

Since 1950, when the Valuation Office of the Inland Revenue (now the VOA) took over responsibility for producing valuation lists from local authorities, shop areas have been zoned for the purposes of rating assessments.

The acceptance by the courts of zoning for such purposes has been almost total, for three basic reasons:

(a)   because zoning reduces rents of let shops to a unit price, and that unit price can be applied to vacant or owner-occupied, etc., shops, thus providing a unit of comparison;
(b)   because there is almost always evidence available to support the unit price; and
(c)   because the method of analysis has 'stood the test of time and experience'.

The Valuation Office has adopted 'arithmetical' or 'arbitrary' zones when analysing and valuing shops for rating purposes.

Although there may be local variations, a typical formula is to adopt 6.1 m (20 feet) for Zones A and B, with Zone C as the remainder; and to 'halve back' rental values.

There are, of course, advantages of simplicity and nation-wide uniformity in adopting 'arithmetical' zones, both for the Valuation Office and for the courts which accept this method.

However, as Emeny and Wilks (1984) state (p. 262):

Such a method cannot be theoretically supported as a method of valuation, it is nothing more than an arithmetical formula which is bound to produce inconsistent results in most cases.

Like zoning, rating assessments are a means to an end and, on the evidence produced by Emeny and Wilks, they should not be considered as valuations or analysis, except for the purposes of rating.

Despite the fact that zoning is used and accepted for rating purposes, there is no requirement that it be the only method used (*Footman, Pretty & Co. v. Chandler (VO)* (1960), p. 23).

In fact the court has rejected zoning in favour of the 'overall price' method in some cases (e.g., *J. Sainsbury Ltd. v. Wood (VO)* (1986)). However, this situation will only arise where:

(a)   ample rental evidence exists to support both the method used and the unit of comparison applied; and
(b)   the method of analysis has been tested in the market.

As yet, with the exception of the 'overall price' method, no other method of valuation has apparently been reported as tested for rating purposes, either in valuation journals or in court proceedings, despite the court's specific mention of such a requirement (*Trevail (VO) v. C & A Modes Ltd.* (1967), p. 205).

Quantity allowance in rating

A quantity allowance becomes important when units of rental value derived from 'small' shops are applied to the valuation of 'large' shops, because the two shop types, as defined by size, are not truly comparable.

'Small' shops and 'large' shops form two distinct markets, because the type of trader who requires the retail area provided by a 'small' shop is not going to be interested in renting a 'large' shop, unless allowed to forgo paying rent for the excess shop area which is not required. In contrast, a trader needing a large retail area will not be interested in a 'small' shop.

In 1956, when a new valuation list took effect, and there was insufficient rental evidence to support the use of an 'overall price' method of valuation (then called an 'intuitive approach'), zoning was applied to 'large' shops and their rating assessments (rental values) fixed by comparison with rents paid for 'small' shops.

At the time, the Lands Tribunal took the view that the effect of size on rents was to be considered in relation to the demand (or lack of demand) for that size, and because of the fact that rents from 'small' shops were used to assess 'large' shops for rates, an end allowance for 'quantity' was generally allowed in the assessments of large shops.

For the 1963, 1973, 1990 and 1995 lists, such an end allowance has had to be justified with market evidence.

With the radical change in retailing patterns, which created positive demand for such 'large' shops as 'departmenta' stores, supermarkets, chain stores, superstores and hypermarkets, there can no longer be a presumption that excessive size in a shop has no value.

Although zoning still applies analysed rents of 'small' shops to the analysed areas of 'large' shops, the courts will require actual market evidence to show that a quantity allowance should be given. Once that market evidence has been produced, it can, of course, be rebutted (*Trevail (VO) v. C & A Modes Ltd.* (1967)).

In any court case, the onus of proof is always on the appellant (the party who has appealed to the court). The appellant should, therefore, always seek to prove either that a quantity allowance would be given in the open market, or alternatively that it would not be given in the market, depending on his point of view.

Market evidence must always be produced in support of either argument.

It should always be remembered that a quantity allowance is only one way of reflecting the discrepancy which may result from the use of 'small' shop rents to value 'large' shops, and that if the market would reflect quantity by another means, an end allowance for quantity would be a duplication.

Metrication

As mentioned earlier, zoning has been applied throughout the UK to the analysis of shop rents and properties for rating purposes since the 1950s. In the 1950s and 60s, all areas of properties would have been calculated using imperial (feet and inches) measurements.

In the early 1970s, metrication was introduced into the UK. However, all the survey details held by the Valuation Office remained in imperial measurements.

Because of the lack of time and resources, the Inland Revenue took a policy decision not to convert all their records to the metric equivalent for the 1973 valuation list. Instead, the imperial measurements were retained, and zoning based on imperial measurements.

Thus, instead of adopting 6 m depths for all shop properties, the valuation office used 6.1 m depths, because 6.1 m is exactly 20 feet. In this way, all survey measurements, areas in terms of Zone A, etc., remained unaltered for the 1973 revaluation and all subsequent alterations to that list.

Because of lack of staff, not all areas were converted to metric equivalents for the 1990 revaluation, and thus the zoning analysis was based on something other than previously existing patterns. This must raise an even greater doubt over the validity of arithmetical zoning and its use for rating purposes.

It is widely recognised in the profession that conversion from imperial (20 feet) to metric (6 m) measurements for zones will alter the value of retail premises. Although the reduction in depth of Zones A and B will be only 0.1 m in each case, when multiplied by the frontage and by a unit price in terms of Zone A, there could be large discrepancies for some units.

It should, however, be added that, like zoning, rating assessments are only a means to an end, i.e., a comparable and uniform basis on which to tax individual occupiers of property.

It is debatable whether the purities of valuation principles have any place in the equity of taxation.

Zoning in court

When a valuation is to be agreed by negotiation, failing which it is subject to the court's determination, it is essential to take into account the attitude of the courts to producing valuations.

To produce a valuation based on one method of valuation for the purposes of negotiation, knowing that such a valuation cannot be justified in court, is potentially dangerous, professionally embarrassing and financially damaging.

The attitude of the judiciary to the zoning method of valuation and the alternatives which have been put forward for the courts' consideration are clearly set out in a number of (mainly rating) cases, most of which relate to, and deal with the problems of, valuing 'large' shops.

Provided that adequate, reliable evidence is presented in support, the courts will accept any method of valuation; but bear in mind that the facts and circumstances of a particular case may render one method of valuation more appropriate than another.

I do not think that, generally speaking, the tribunal regards any one method as inherently better than another, provided that the system which is adopted is applied by a practitioner who is thoroughly familiar with the processes involved, is properly and accurately informed, and can support his figures with adequate statistical information. (*Footman, Pretty & Co. v. Chandler (VO)* (1960), p. 23.)

It is this overriding need of the Lands Tribunal to have a method of valuation supported by evidence of comparable values (rental or assessment) which has resulted in the rejection of a 'spot' valuation or intuitive approach (ibid.), overall rental value (*U.D.S. Tailoring Ltd. v. B.L. Holdings Ltd.* (1981)), and turnover of the business (ibid.).

In assessing the rental value of any property, the best evidence is rental evidence, either direct rental evidence of the subject property itself, or indirect rental evidence of comparable property in the near vicinity of the subject property.

If there are an adequate number of rents of large shops, it may be possible ... to value them on a basis deduced from those rents and without resort to the zoning method. (*Trevail (VO) v. C & A Modes Ltd* (1967), p. 199.)

In *F. W. Woolworth & Co. Ltd. v. Moore (VO)* (1978), the Lands Tribunal rejected the zoning method in favour of an overall rental value. The Tribunal found this approach more reliable because there were sufficient actual rents available for larger shops, supported both by comparable assessments (to show the valuation officer's opinion of how the rental evidence was to be adjusted) and by the actual negotiations for the lease carried out by the parties involved. Such evidence was more reliable than the evidence produced by zoning smaller shops; although the Tribunal did state:

... with regard to the rents and assessments of the smaller shops, I do not go so far as to say that this evidence is irrelevant; but in my opinion it would require some glaring inconsistency to arise before the evidence of the rents and assessments of the small shops could be taken to outweigh the evidence of the actual rents for the large shops. (p. 215)

In *J. Sainsbury Ltd. v. Wood (VO)* (1980), where there was no market rental evidence for any of the comparable properties, the Tribunal said that the rating assessments of the comparable properties: 'assume the character of actual rents properly adjusted in terms of [rateable] value' (p. 210) and relied on comparable assessments to provide a rental value in terms of Zone A.

Under the rules of natural justice, a court can only base a decision on evidence given to and examined by it. If evidence is not presented, it cannot be presumed. It is, therefore, essential that relevant, reliable and, hopefully, conclusive evidence is presented by each party to support its arguments to the court.

The court will rely most heavily on the evidence it finds most relevant and reliable in the open market.

The courts recognise that '[t]he zoning system is but a means to an end' (*B.H.S. Ltd v. C.B. of Brighton & Burton* (VO) (1958), p. 350) and will expect the 'system' to be adapted to fit the facts of a case and any supporting market evidence.

... although this [zoning] is a proper approach particular circumstances may require modification either upwards or downwards of the basic rent thus arrived at. (*Janes (Gowns) Ltd. v. Harlow Development Corporation* (1980), p. 799.)

In the above case, which involved determination of a rent and other terms in a lease granted under Part II of the Landlord and Tenant Act 1954, regard was had to user covenants in the lease.

... the relevant facts must always be the dominant factor in arriving at correct assessments in each case and those facts may well justify modifications in the application of a particular method of valuation. (*Trevail (VO) v. C & A Modes Ltd.* (1967), pp. 198–9.)

It follows, therefore, that while the courts will accept the arithmetical or arbitrary method of zoning, they recognise it as a mathematical exercise and will thus look beyond it to see whether its use produces a reasonable rental value.

Justification for zoning

The courts recognise and accept that zoning is based on the principle that it is the frontage of shop property in a busy high street which attracts customers into the shop, and, that because of 'its proximity to the bustle of the city pavements and the inducement offered to window shoppers' (*J. Sainsbury Ltd. v. Wood (VO)* (1980) p. 211), the front part of a shop is more valuable than the rear.

Where, however, there is no pavement trade, as in the case of an out-of-town superstore, where most customers arrive by car, the Lands Tribunal (in *J. Sainsbury Ltd. v. Wood (VO)* (1986)) found it impossible to say that any particular part of the sales area was more valuable than any other part and, in such circumstances, rejected the zoning method in favour of an overall price per square [metre] (which was supported by evidence of agreed rating assessments).

Even where shops are located within the high street, other criteria, which reflect the changes in retailing patterns, may affect the suitability of the zoning method in any given case.

Nowadays it is common to find 'walk-round' or 'departmental' stores which are ten times or even fifteen times as large as the 'normal' shop in a particular centre. Again, during the last decade or so, the number of 'chain stores' and the number of stores each company operates has increased enormously with the result that there is potentially greater competition for suitable accommodation. ... It can be argued that ... there comes a point when the size of a store takes it right out of the class of ordinary shops ... (*Trevail (VO) v. C & A Modes Ltd.* (1967), p. 203–5.)

... I consider that the application of a zone 'A' price derived from relatively small shops to the valuation of large walk-round or departmental stores is a pure distortion of what in other circumstances is an excellent method of valuation. (*B.H.S. Ltd v. C.B. of Brighton & Burton (VO)* (1958) p. 348.)

and later,

To apply the zone value derived from these shops where the size is in the region of 1,500 sq. ft. (140m²) to the large walk-round and departmental stores seems ... to be stretching the limits of this method of valuation a good deal beyond its capabilities. (ibid.)

The complexities of the calculations necessary if one is to try to find the value of a store with over 43,000 sq. ft. [4,000 m²] of floor space from the rental evidence relating to shops with an area of only 2,000 sq. ft. [185 sq. m.] or 3,000 sq. ft. [280 m²] and the consequences which follow from a variation in price or other valuation details incline us to the view that some other less complicated method of valuation of 'large' shops or stores should be tested. (*Trevail (VO) v. C & A Modes Ltd.* (1967) p. 204.)

However, whether or not there is adequate evidence to support other methods of valuation, the zoning method can be justified:

... where the zoning method is properly based upon local values which are established or agreed, it is susceptible of test by precise comparison, and if applied by an experienced practitioner and confined to the type of hereditament for which it is suitable, more appropriate for submission to the tribunal. (*Footman, Pretty & Co. v. Chandler (VO)* (1960) p. 26.)

– and later:

Put simply, I consider that a zoning method of valuing shop premises provides reasons and results which can be examined and tested, that, properly and consistently used, it is the best method of valuation which has yet been devised, and that it could be used with advantage in all cases for which it is suitable. (ibid. p.27.)

It is the failure of the valuation profession to test any less complicated method of valuation for 'large' shops which has resulted in the continued use of zoning for properties to which its principles do not apply.

Where there is no rental evidence available from comparable properties, the zoning of smaller 'normal' shops can be used to provide a rental value for 'large' shops. (This is one of the stated objectives of the zoning method.)

However, the courts recognise that it is inappropriate to apply the zoning method to 'large' shops, and will do so only in the absence of better evidence, i.e., rental values of comparable properties.

... the zoning method was the only method of valuation advanced at the hearing, we have therefore had no alternative but to adopt that method ... (*Trevail (VO) v. C & A Modes Ltd.* (1967) p. 204.)

In *U.D.S. Tailoring Ltd. v. B.L. Holdings Ltd.* (1981), where evidence of rental value was given based on the turnover of business and the overall average rental per square foot, the court preferred to rely on the evidence which supported the zoning method and added:

... but I must emphasise that I have reached that conclusion on the evidence adduced before me and I am not approaching it in any way as a matter of principle.

Summary

- The best evidence on which to base a rental valuation is comparable rental evidence.
- Any method of valuation must be supported by adequate, reliable evidence of rental value, whether direct or indirect, and for rating valuations, comparable assessments.
- Despite the fact that there is no right or wrong method of valuation for any particular type of property, the zoning method of valuation is accepted in the calculation of a rental value for 'normal' shops.
- In the absence of supporting rental evidence, the zoning method can be adapted to provide a rental value of 'large' shops based on the rental evidence of 'normal' shops.
- It is necessary to rely on other alternative techniques being tested and accepted by practitioners in the market and before the courts, if zoning is to be relegated to use in only appropriate circumstances, or abandoned altogether.

Appendix F – Criteria for a tax

Any tax can be viewed from the point of view of the various groups involved in the system and should reflect the concerns of all parties, as well as certain general criteria.

Criteria for the taxpayer

A tax should:

1. be simple to understand and in fact be understood;
2. be administered at minimum cost and inconvenience;
3. be payable at convenient times and by a choice of convenient methods;
4. not involve any breach of confidentiality in a taxpayer's personal affairs;
5. not suffer from illogical, unfair or historical exemptions;
6. have a basis which is up-to-date;
7. not be arbitrary;
8. be seen to be spent efficiently and wisely on socially-acceptable services;
9. have accessible, cheap, reliable and speedy methods of appeal;
10. be seen to ensure an equitable burden of tax liability between taxpayers of different means and assets (horizontal and vertical equity);
11. provide rebates to reflect current views on liability for those of limited means; and
12. not be required to raise an excessive amount of revenue.

Criteria for the spending authority

A tax should:

1. be unique to the spending authority;

2.   apply universally on a common and uniform basis;

3.   be determined by one authority without interference from any other authority;

4.   relate wholly and demonstrably to the spending authority and its jurisdiction;

5.   have a yield which is certain, predictable and adequate;

6.   be stable but adjust rapidly to changing economic circumstances;

7.   not conflict with contemporary views on the roles and responsibilities of the spending authority;

8.   be adequate to maintain or expand, in accordance with current policies, the range and level of services; and

9.   not be overstretched to meet the financial demands of the spending authority beyond its inherent capabilities.

Criteria for the tax collector

A tax should:

1.   be difficult to evade and avoid;

2.   be contained in legislation which is mandatory, clear and unambiguous and which is not dependent on extensive litigation for interpretation;

3.   be collected and recovered by simple, cheap and fair procedures;

4.   be easily explained to taxpayers; and

5.   involve minimal losses and leakages.

Generally

A tax should:

1.   not impede central management of the economy;

2.   not conflict with any other taxes nor bring the various government tiers or agencies into competition or conflict;

3.   be capable of operating within the existing framework of government;

4.   not require another committee of inquiry for at least 25 years;

5.   not be introduced without full regard for all the implications;

6.   promote full democratic accountability; and

7.   clearly state who does what for whom, at what cost and at whose expense.

(adapted from C. Stuart Page, 'Local Taxation – a preliminary check list for Layfield', *Rating and Valuation*, June 1976, p. 163).

Accountability

The issue of democratic accountability is an interesting one. It figured prominently in the design of the Uniform Business Rate, the Community Charge (the poll tax) and the Council Tax. Local authorities are required to present to all taxpayers details of how their revenue is spent and this is normally achieved by an enclosure sent together with the tax demand.

However, this is not a principle respected by central government. Such a show of double standards is curious, with central government's use of 'democratic accountability' as one of the justifications for the fact that the UBR is now a tax levied by central government, effectively, on behalf of local government.

Appendix G – Glossary

annual equivalent the rental equivalent of a capital sum (normally called a premium) arrived at by dividing the capital sum by the appropriate Years' Purchase

antecedent valuation date a date fixed in advance of the rating list coming into effect which is the date at which all hereditaments are valued, e.g., for the rating list which came into effect on 1 April 1995, the antecedent valuation date is 1 April 1993

assessment the rateable value of a hereditament

billing authority a local authority empowered to collect the Uniform Business Rate and the Council Tax

capping central government's power to limit local authorities' spending and thereby to limit their ability to raise finance through the Council Tax

central rating lists a list of hereditaments compiled under s. 52 of the 1998 Act containing entries for hereditaments occupied by canal, electricity, gas, railway, telecommunication, water supply and long-distance pipelines occupiers (see 8.4)

certificate of value a certificate provided by the valuation officer establishing the rateable value of a hereditament at a given time or of a portion of a hereditament on which rates are payable for a specified time under Part VI, Non Domestic Rating (Chargeable Amounts) Regulations 1994

completion notice a notice served by a billing authority on an owner of a building which is or which, in the opinion of the billing authority, will be complete within three months, under s. 46A and Sch. 4A, 1988 Act (see 2.4.28–46)

composite hereditament a hereditament part of which comprises domestic property, defined in s. 64 (9) 1988 Act.

dead frontage frontages which do not involve trade and tend to present psychological barriers to pedestrian shoppers and inhibit them from proceeding further, such as hoardings, dwellings and (often) public houses

decapitalisation rate the factor applied to a capital value to convert it into a rental value within a contractors' test valuation

derating the total or partial exemption of a class of property from rate liability

distraint or distress the seizure of chattels by an agent of a billing authority in order to recover arrears of rates

domestic property property used wholly for the purpose of living accommodation, defined in s. 66 (1) 1988 Act.

effective capital value (ECV) the sum of the capital cost of constructing a building plus the value of the land, reduced by any allowance for depreciation, used in the contractors' test

effective date date on which an alteration to a rating list takes effect

empty rate the rate levied on the owner of an empty hereditament – the same as unoccupied rate

end allowance a deduction made at the end of a valuation to reflect factors (normally disabilities) which have not yet been taken into account in the valuation

equation theory equation theory assumes that an occupier has a given amount of money out of which to pay both rent and rates. The occupier will be ambivalent as to how that money is split between rent and rates, provided the overall sum is not exceeded. (see Emeny and Wilks, 1984, pp. 187–90).

hereditament strictly, property which is, or may become liable to a rate, being a unit of such property which is or would fall to be shown as a separate item in the rating list (s. 115 General Rate Act 1967). One hereditament has one rateable value and one entry in the rating list.

hypothetical tenancy the relationship assumed between the tenant mentioned in the definition of rateable value and his implied landlord

hypothetical tenant   the tenant referred to in the definition of rateable value

interested person   an occupier, a person (other than a mortgagee not in possession) having in any part of the hereditament either a legal or equitable interest that would entitle him to possession (after the cessation of any prior interest); and any person having a 'qualifying connection' with the occupier or other interested person. A 'qualifying connection' interest reflects the legal connection which may exist between, e.g., a parent company and its subsidiary. Interested persons are entitled to make proposals to alter the rating list.

listing officer   the valuation officer (refer), renamed for Council Tax purposes

local non-domestic rating list   a list of non-domestic hereditaments situated within the area of a billing authority, compiled and maintained by the valuation officer for that billing authority and which takes effect every five years (currently from 1 April 1995)

overbid   the additional money paid by a potential tenant or purchaser in excess of market value, generally because of the particular need of the tenant/ purchaser for the property, e.g., local authorities may offer an overbid to secure premises necessary to undertake a statutory duty

premium   the capital sum paid by an actual or prospective tenant to a landlord, usually in return for a reduction in the amount of rent otherwise payable

proposal to alter the list   the procedure under which an interested person or relevant authority seeks an alteration of the rating list

rateable occupation   actual possession, exclusive occupation, beneficial occupation and a sufficient degree of permanence which give rise to occupied rate liability (see 2.3).

rateable value   the rent at which it is estimated the hereditament might reasonably be expected to be let from year to year if the tenant undertook to pay all usual tenant's rates and taxes and to bear the cost of the repairs and insurance and the other expenses (if any) necessary to maintain the hereditament in a state to command that rent (defined in Sch. 6 para.

2 (1) 1988 Act) – basically it is a net annual rental value (see 5.3)

rating lists either the local non-domestic rating list or the central non-domestic rating list (see above)

rebus sic stantibus 'things as they stand'; used to refer to the circumstances in which a hereditament is valued, which means that no physical changes or changes of use can be assumed

revaluation the reassessment of all hereditaments by the Valuation Office Agency and the compilation of new rating lists, which occurs every five years (1990 and 1995)

tenant's share a combination of risk, remuneration and interest on capital invested, required by a tenant before an amount for rent and rates can be ascertained within a profits method of valuation.

tone of the list (or tone) the level of rental value as at the antecedent valuation date, which is applied to the valuation of all hereditaments during the life of the rating list

Valuation Office Agency the privatised organisation of valuation officers, appointed by the Commissioners of Inland Revenue, responsible for the compilation and maintenance of rating lists and valuation lists

valuation officer the officer, appointed by the Commissioners of Inland Revenue, responsible for the compilation and maintenance of the rating list for the area covered by the billing authority.

valuation tribunal the court of the first instance for appeals against entries in the rating lists, completion notices, certificates of value, Council Tax banding, etc.

Year's Purchase (YP) a capitalisation factor, used to covert a rental value into its capital equivalent and *vice versa*. It is also called the present value of £1 per annum.

zoning a method of analysing shop areas in order to compare the rental value of one shop property with that of another.

References and Bibliography

Anon, 1995. *Rent-Free Periods and Valuation.* (1995) 19 *EG* 143–4.

Baum, Andrew and Mackmin, David. 1997. *The Income Approach to Property Valuation.* Routledge & Kegan Paul.

Britton, William; Davies, Keith and Johnson, Tony. 1989. *Modern Methods of Valuation.* The Estates Gazette Ltd.

Brownlow, G. S., 1947. *Site Value of Shops.* The Estates Gazette Ltd.

Colborne, Ann and Hall, Philip, C. L., 1992. 'The profits method of valuation'. *Journal of Property Valuation & Investment* Vol. 11 No. 1. Autumn 1992.

Colborne, Ann, 1992. 'The profits basis of valuation'. Paper presented at Valuation Techniques Seminar, The Royal Institution of Chartered Surveyors, 13 March 1992.

Council Tax: Practice Note No. 1 – valuation lists (revised August 1993) (prepared by the Department of the Environment, the Welsh Office, and associations of local authorities).

Council Tax: Practice Note No. 2 – Liability, discounts and exemptions (revised July 1993) (prepared by the Department of the Environment, the Welsh Office, and associations of local authorities).

Council Tax: Practice Note No. 3 – Council Tax Benefit (prepared by the Department of the Environment, the Welsh Office, and associations of local authorities).

Council Tax: Practice Note No. 4 – Transitional Arrangements (revised February 1993) (prepared by the Department of the Environment, the Welsh Office, and associations of local authorities)

Council Tax: Practice Note No. 5 – Administration (including billing and collection) (revised April 1994) (prepared by the Department of the Environment, the Welsh Office, and associations of local authorities).

Council Tax: Practice Note No. 6 – Appeals (prepared by the

Department of the Environment, the Welsh Office, and associations of local authorities).

Council Tax: Practice Note No. 7 – Tax setting, precepting and levying (Revised August 1993 and August 1995) (prepared by the Department of the Environment, the Welsh Office, and associations of local authorities).

Council Tax: Practice Note No. 9 – Recovery and Enforcement (revised September 1993) (prepared by the Department of the Environment, the Welsh Office, and associations of local authorities).

Davies, Keith, 1994. *Law of Compulsory Purchase and Compensation.* Butterworths.

Emeny, Roger and Wilks, Hector, 1984 *Principles and Practice of Rating Valuation.* 4th edition The Estates Gazette Ltd.

Gronow, Stuart and Plimmer, Frances, 1987. 'Zoning: A Nonsense for the 21st Century' *Rating and Valuation* p. 107.

HMSO, 1976. *Local Government Finance: Report of the Committee of Inquiry.* HMSO. Cmnd 6453 (The Layfield Report).

HMSO, 1981. *Alternatives to Domestic Rates.* HMSO. Cmnd 8449.

HMSO, 1983. *Rates – proposals for rate limitation and reform of the rating system.* HMSO.

HMSO, 1986. *Paying for Local Government.* HMSO. Cmnd 9714.

HMSO, 1991. *A New Tax for Local Government.* HMSO.

HMSO, 1993. *Rating of plant and machinery: a report by the Wood Committee.* HMSO March 1993. Cmnd 2170.

Jones, Long Wootton, 1989, *The glossary of property terms.* The Estates Gazette Ltd.

Marshall, Harvey and Williamson, Hazel, 1997. *Law and Valuation of Leisure Property.* (2nd edn.) The Estates Gazette Ltd.

Millington, Alan, 1994. *An Introduction to Property Valuation.* The Estates Gazette Ltd.

Page, C. Stuart, 1979. Local taxation – a preliminary check list for Layfield'. *Rating and Valuation* June 1979, p. 163.

Page, C. Stuart, 1980. 'Rates – the right tax for local government'. *Rating and Valuation Reporter* October 1980, pp. 229–34.

Rating and Valuation Association (now the Institute of Revenue and Rating Valuation), 1983. *Shops – Valuation for Rating: Report of a Research Group.* Rating and Valuation Association.

Rees, W. H. (ed.). 1992. *Valuation: Principles into Practice.* The Estates Gazette Ltd.

Richmond, David, 1981. *Introduction to Valuation.* The Estates Gazette Ltd.

RICS, 1996. *Improving the System.* The National Committee on

Rating, The Royal Institution of Chartered Surveyors. (The Bayliss Report).

RICS/ISVA, 1993. *Code of Measuring Practice.* The Royal Institution of Chartered Surveyors and the Incorporated Society of Valuers and Auctioneers.

Rose, Jack, 1988. *Rates vs Poll Tax* [1988] 03 EG 81 & 85

Sales, Harry, B., (Gen. Ed.) 1997. *Encyclopedia of Rating and Local Taxation.* Sweet & Maxwell

Scorrett, Douglas, 1991. *Property valuation: the five methods.* E & FN Spon.

Sedgwick, J. R. E., 1992. Garages and Petrol Stations, in Rees, W. H., (ed.) *Valuation: Principles into Practice* (4th edn) The Estates Gazette Ltd. pp. 425–59.

Smith, Kenneth, 1984. *A Consideration of the Zoning Principle.* 270 *EG* 388.

Stewart, Chris and Reynolds, Steven, 1993. *A Guide to the Council Tax.* Tolley Publishing Company.

Ward, Martin, 1993. *Council Tax Handbook.* Child Poverty Action Group.

Widdicombe, D. Eve, Trustram, D. and Anderson A., 1976. *Ryde on Rating.* Butterworths.

Wilks, Hector. 1985. *A Case for the Present System.* 275 *EG* 26 & 28.

Wilks, Hector, 1986. *Some Further Thoughts.* 277 *EG* 936.

Wilks, Hector, 1987. 'Property tax systems'. *The Valuer* January/February 1987, p. 3.

Index